Intermetallic Compound

Intermetallic Compound

Editor

Jacek Ćwik

MDPI • Basel • Beijing • Wuhan • Barcelona • Belgrade • Manchester • Tokyo • Cluj • Tianjin

Editor
Jacek Ćwik
Institute of Low Temperature
and Structure Research,
Polish Academy of Sciences
Poland

Editorial Office
MDPI
St. Alban-Anlage 66
4052 Basel, Switzerland

This is a reprint of articles from the Special Issue published online in the open access journal *Crystals* (ISSN 2073-4352) (available at: https://www.mdpi.com/journal/crystals/special_issues/Intermetallic_Compound).

For citation purposes, cite each article independently as indicated on the article page online and as indicated below:

LastName, A.A.; LastName, B.B.; LastName, C.C. Article Title. *Journal Name* **Year**, *Volume Number*, Page Range.

ISBN 978-3-0365-0138-3 (Hbk)
ISBN 978-3-0365-0139-0 (PDF)

Contents

About the Editor

Jacek Ćwik from 2018 is acting deputy branch manager of the Division of Low Temperature and Superconductivity at the Institute of Low Temperature and Structure Research, Polish Academy of Sciences, in Wroclaw. He started his scientific career in 2000 after graduating in Material Engineering at the Technical University of Wroclaw, Department for Basic Problems of Technology, in Poland. After graduation, he worked as an Assistant at the International Laboratory of High Magnetic Fields and Low Temperatures, Wroclaw. In 2006, he defended his doctoral dissertation—Institute of Inorganic Chemistry and Rare Earth Elements, Wroclaw, Poland—thesis: "Structural and physical properties of HoNi2 – MNi2 , (M = Sc, Y, La, Lu)". In 2012, he became the leader of the group at the International Laboratory of High Magnetic Fields and Low Temperatures involved in the study of materials with significant effects induced by the magnetic field. In 2016, he obtained his habilitation in Physics Sciences at the Institute of Molecular Physics, Polish Academy of Sciences, Poznan, Poland—thesis "The influence of modifying composition on magnetic and magnetocaloric properties for chosen solid solutions of Laves phase structure". Since 2017, he has been working as an Assistant Professor at the Institute of Low Temperature and Structure Research, Polish Academy of Sciences, in Wroclaw. His scientific interests are focused on the fabrication of intermetallic compounds with specific magnetic properties via the arc melting method, as well as measurements of their thermomagnetic and thermodynamic properties. He is the co-author of a unique position for direct measurements of the magnetocaloric effect using high magnetic fields generated by resistance Bitter type-magnets.

Preface to "Intermetallic Compound"

In modern society, applications of new functional materials are diverse and countless. Intermetallics, namely, compounds formed by two or more metallic elements, are among the various novel and significant material developments. By definition, intermetallics are compounds of metals whose crystal structures are different from those of the constituent metals; thus, intermetallic phases and ordered alloys are included. Due to both their specific physical and mechanical properties, they occupy an intermediate position between metal and non-metal. Notably, intermetallic compounds have become a powerful spectrum for research in various branches of science. The enormous interest of scientists has contributed to the creation of various interesting compounds. In recent decades, a great number of scientific books and articles have been written, and many scientific meetings, such as conferences and workshops on intermetallic compounds, have been held. Intermetallics have received considerable attention, owing to their potential for various applications, such as their use as permanent magnets, magneto-optical recording media, hydrogen storage and magnetocaloric materials, turbine blades, and in microelectronics. With a favorable combination of high strength, low density, and good corrosion resistance, intermetallics are specifically suited for applications at high temperatures and in adverse environments. They can also display desirable magnetic, superconducting, and chemical properties due to their strong internal order and mixed metallic and covalent or ionic bonding, respectively.

Jacek Ćwik
Editor

Editorial

Special Issue "Intermetallic Compound"

Jacek Ćwik

Institute of Low Temperature and Structure Research, Polish Academy of Sciences, Okólna 2, 50-422 Wrocław, Poland; j.cwik@intibs.pl

Received: 18 November 2020; Accepted: 19 November 2020; Published: 23 November 2020

This Special Issue collects ten articles related to the broadly understood physical properties of intermetallic compounds. The most important results of individual articles are presented below.

In the first article [1], the authors present experimental studies on the reduction of neodymium from a fluoride–chloride salt mixture, KCl–NaCl–$CaCl_2$–NdF_3, melted in a shaft electric furnace at different temperatures. The influence of technological parameters on the degree of extraction of neodymium was evaluated. It was experimentally proven that when zinc is added to a reducing agent (magnesium), the degree of extraction of neodymium into the master alloy is 99.5–99.7%. Analysis of the microstructure showed that the obtained ternary master alloy, 25Mg–50Zn–25Nd, is characterized by a uniform distribution of intermetallic compounds (Mg_3NdZn_6) in the bulk of the double magnesium–zinc eutectic. The obtained experimental data are a prerequisite for the development of industrial technology for the production of magnesium–zinc–neodymium alloys for their use in non-ferrous and ferrous metallurgy.

Roni Z. Shneck et al. [2] present detailed research on $MgTiO_3$ formed at low temperatures by the sol–gel synthesis technique. Accurate thermal expansion coefficients were measured for sol–gel products of stoichiometric $MgTiO_3$. The lattice parameters of $MgTiO_3$ made by sol–gel synthesis measured in High-temperature X-ray diffraction (HT-XRD) between 25 and 890 °C are well integrated with the previously reported HT-XRD study of sol–gel $MgTiO_3$ product between 700 and 1300 °C. The lattice parameters of stoichiometric $MgTiO_3$ sol–gel products are slightly lower than nonstoichiometric $MgTiO_3$ with maximal excess of Ti.

Songbai Xue et al. presented three articles [3–5]. These articles were related to the determination of the effects of the addition of various nanoparticles (GaF_3, ZnF_2, $Zn(BF_4)_2$ and Ga_2O_3) on the activity of CsF-RbF-AlF_3 flux and the mechanical behavior of Al/steel brazed joints. Among others, it was shown that the CsF-RbF-AlF_3 flux doped with GaF_3 obtained the maximum spreading area, both on 6061 aluminum alloy and Q235 low-carbon steel, and reached the highest shear strength of 126 MPa.

Nur Baizura Mohamed et al. [6] have studied the effect of Bi substitution on the structural and magnetic susceptibility properties of $Nd_{1-x}Bi_xMnO_3$. The XRD analysis and Rietveld refinement show the pattern corresponding to the perovskite-type $NdMnO_3$, which crystallizes in the orthorhombic system with main peak at (121) hkl plane with the Pbnm space group. The samples show a strong antiferromagnetic to paramagnetic transition existing at 76 K, 77 K and 67 K for samples (x = 0, 0.25 and 0.50), respectively. Deviation of temperature dependence of inverse susceptibility curves shows the existence of Griffiths phase in these materials.

In the next article, Natalia Kolchugina et al. [7] present research related to simulating the hysteretic characteristics of hard magnetic materials based on $Nd_2Fe_{14}B$ and $Ce_2Fe_{14}B$ intermetallics. The simulation and analysis of hysteresis loops of the quasi-ternary intermetallics $(Nd_{1-x}Ce_x)_2Fe_{14}B$ (x = 0–1) was performed. Results of the simulation indicate that the alloying of the $Nd_2Fe_{14}B$ intermetallic with Ce to x = 0.94 does not completely eliminate the negative effect of spin-reorientation phase transition on the residual magnetization of the $(Nd_{1-x}Ce_x)_2Fe_{14}B$ intermetallic and slightly decreases the slope of magnetization reversal curve.

In the next work [8], the authors develop a grid search (GS)-based extreme learning machine (ELM) and hybrid gravitational search algorithm (GSA)-based support vector regression (SVR) for estimating the maximum magnetic entropy change (MMEC) of doped manganite-based compounds. The precision of the developed models offers a green solution to the known pollutant-based refrigerants and presents ways by which manganite-based compounds of the desired maximum magnetic entropy change can be predicted for possible laboratory fabrication and implementation.

Yao Jiang et al. [9] present work related to reactively synthesized porous Ti_3SiC_2 compound and its mechanical properties with different apertures. The flexural stress–strain curves of porous Ti_3SiC_2 compounds exhibit two stages of elastic deformation and fracture, and the elastic deformation behavior shows a characteristic of kinking nonlinear elastic solid. The reduction of the pore size of porous Ti_3SiC_2 improves the phase purity, strength and stiffness while increasing its filtration accuracy.

In the last article, Qian Gao et al. [10] present an experimental study on Zn-doped Al-rich alloys for fast on-board hydrogen production. The hypothesis that Zn will change the hydrogen production performance of the alloy by entering the GB phase was verified and it was found that the reaction mechanism cannot only be explained by the eutectic reaction of the GB phase and Al.

I hope that the presented set of articles will arouse genuine interest among readers and, perhaps, encourage them to carry out their own successful research in the field of intermetallic compounds.

Funding: This research was funded by the National Science Center, grant number 2019/33/B/ST5/01853.

Acknowledgments: Thanks to all contributing authors of this Special Issue and the editorial staff of Crystals.

Conflicts of Interest: The authors declare no conflict of interest. The funders had no role in the design of the study; in the collection, analyses, or interpretation of data; in the writing of the manuscript, or in the decision to publish the results.

References

1. Beloglazov, I.; Savchenkov, S.; Bazhin, V.; Kawalla, R. Synthesis of Mg–Zn–Nd Master Alloy in Metallothermic Reduction of Neodymium from Fluoride–Chloride Melt. *Crystals* **2020**, *10*, 985. [CrossRef]
2. Tuval, T.; Rosen, B.A.; Zabicky, J.; Kimmel, G.; Dilman, H.; Shneck, R.Z. Thermal Expansion of $MgTiO_3$ Made by Sol-Gel Technique at Temperature Range 25–890 °C. *Crystals* **2020**, *10*, 887. [CrossRef]
3. Yao, Z.; Xue, S.B.; Zhang, J.X. Effect of Various Nanoparticles (GaF_3, ZnF_2, $Zn(BF_4)_2$ and Ga_2O_3) Additions on the Activity of CsF-RbF-AlF_3 Flux and Mechanical Behavior of Al/Steel Brazed Joints. *Crystals* **2020**, *10*, 683. [CrossRef]
4. Yao, Z.; Xue, S.B.; Zhang, J.X. Comparative Study on the Activity of GaF_3 and Ga_2O_3 Nanoparticle-Doped CsF-AlF_3 Flux for Brazing 6061 Al/Q235 Steel Joints. *Crystals* **2020**, *10*, 498. [CrossRef]
5. Yao, Z.; Xue, S.B.; Yang, J.L.; Zhang, J.X. Inducing the Effect of a Ga_2O_3 Nano-Particle on the CsF-RbF-AlF_3 Flux for Brazing Aluminum to Carbon Steels. *Crystals* **2020**, *10*, 183. [CrossRef]
6. Azhar, N.A.; Ismail, I.S.; Mohamed, N.B.; Hashim, A.; Mohamed, Z. Effect of Bi Substitution on Structural and AC Magnetic Susceptibility Properties of $Nd_{1-x}Bi_xMnO_3$. *Crystals* **2020**, *10*, 521. [CrossRef]
7. Kolchugina, N.B.; Zheleznyi, M.V.; Savchenko, A.G.; Menushenkov, V.P.; Burkhanov, G.S.; Koshkid'Ko, Y.S.; Ćwik, J.; Dormidontov, N.A.; Skotnicova, K.; Kursa, M.; et al. Simulating the Hysteretic Characteristics of Hard Magnetic Materials Based on $Nd_2Fe_{14}B$ and $Ce_2Fe_{14}B$ Intermetallics. *Crystals* **2020**, *10*, 518. [CrossRef]
8. Ibn Shamsah, S.M.; Owolabi, T.O. Modeling the Maximum Magnetic Entropy Change of Doped Manganite Using a Grid Search-Based Extreme Learning Machine and Hybrid Gravitational Search-Based Support Vector Regression. *Crystals* **2020**, *10*, 310. [CrossRef]
9. Jiang, Y.; Liu, X.L.; Gao, H.Y.; He, Y.H. Reactively Synthesized Porous Ti_3SiC_2 Compound and Its Mechanical Properties with Different Apertures. *Crystals* **2020**, *10*, 82. [CrossRef]

10. Liu, D.; Gao, Q.; An, Q.; Wang, H.C.; Wei, J.L.; Wei, C.D. Experimental Study on Zn-Doped Al-Rich Alloys for Fast on-Board Hydrogen Production. *Crystals* **2020**, *10*, 167. [CrossRef]

Publisher's Note: MDPI stays neutral with regard to jurisdictional claims in published maps and institutional affiliations.

Article

Synthesis of Mg–Zn–Nd Master Alloy in Metallothermic Reduction of Neodymium from Fluoride–Chloride Melt

Ilia Beloglazov [1,*], Sergey Savchenkov [2], Vladimir Bazhin [3] and Rudolf Kawalla [4]

1. The Automation of Technological Processes and Production Department, Saint Petersburg Mining University, 199106 Saint Petersburg, Russia
2. Patent and Licensing Department, Saint-Petersburg Mining University, 199106 Saint Petersburg, Russia; savchenkov_sa@pers.spmi.ru
3. The Automation of Technological Processes and Production Department, Saint-Petersburg Mining University, 199106 Saint Petersburg, Russia; bazhin_vyu@pers.spmi.ru
4. Research of Technische Universität Bergakademie Freiberg, 09599 Freiberg, Germany; Rudolf.Kawalla@imf.tu-freiberg.de

* Correspondence: beloglazov_ii@pers.spmi.ru; Tel.: +7-921-7785984

Received: 26 September 2020; Accepted: 29 October 2020; Published: 30 October 2020

Abstract: In the presented article, a differential thermal analysis was carried out and the temperatures of thermal effects were established that arise during the reduction of neodymium from a technological salt mixture KCl–NaCl–$CaCl_2$–NdF_3 with a magnesium–zinc alloy. The results of experimental studies on the reduction of neodymium from a fluoride–chloride melt in a shaft electric furnace at temperatures of 550, 600, 650, 700 °C are presented. In order to increase the degree of extraction of neodymium into the Mg–Zn–Nd master alloy, the study of the influence of technological parameters on the degree of extraction of neodymium was carried out. It was experimentally proven that when zinc is added to a reducing agent (magnesium), the degree of extraction of neodymium into the master alloy is 99.5–99.7%. The structure of the obtained master alloy samples, characterized by a uniform distribution of ternary intermetallic compounds ($Mg_{3,4}NdZn_7$) in the volume of a double magnesium–zinc eutectic, was studied by optical and electron microscopy.

Keywords: magnesium–zinc–neodymium master alloy; Mg–Zn–Nd; magnesium master alloy; magnesium; rare-earth metals; master alloy synthesis; metallothermic reduction

1. Introduction

The addition of neodymium into the magnesium–zinc–zirconium system leads to a change in the phase composition of magnesium-based alloys. In this case, simultaneously with the known strengthening phases of magnesium with neodymium (for example, $Mg_{12}Nd$), phases of a more complex stoichiometric composition are formed, containing magnesium, neodymium, and zinc, for example $Nd_{15}Mg_{65}Zn_{20}$, $Nd_{16}Mg_{37}Zn_{47}$, $Nd_6Mg_{41}Zn_{53}$ [1–7]. Due to the presence of these phases in the system, a combination of highly plastic and structural characteristics of magnesium alloys, as well as their high-temperature strength, is ensured, since the thermal expansion coefficient of long period stacking ordered (LPSO) phases is much lower than that of the basic α-solid solution [8–11].

It is known that the production of magnesium and aluminum alloys is carried out by dissolving in the melt double and ternary master alloys, which are produced by fusing the components, using metallothermic reduction of alloying components from their compounds (for example, oxides or fluorides) or electrolysis [12–15]. In recent years, there has been a tendency to produce more complex master alloys, for example, ternary or quadruple [16–18]. It was found that ternary master alloys can be effectively used in the production of alloys based on light metals [19,20].

In this regard, it seems relevant to develop new solutions for the production of complex magnesium master alloys including rare earth metals (REEs). It should be noted that for the domestic magnesium industry the task of obtaining magnesium alloys is of great importance in connection with the approval of the Strategy for the development of the metallurgical industry in Russia for the period up to 2030. In accordance with the sustainable development plan, the task of increasing the production of metallurgical products with high added value was approved, which should lead to a reduction in imports of such products [21]. In addition, the state's priority is to improve the quality of production of Russian metallurgical companies in order to increase the efficiency of processing mineral raw materials [22–29].

The aim of the work is to develop technological solutions that ensure high extraction of neodymium in the ternary Mg–Zn–Nd master alloy during metallothermic reduction of neodymium fluoride, which can be used in the production of high-strength and heat-resistant magnesium and aluminum alloys.

2. Materials and Methods

Differential thermal analysis (DTA) was carried out using a (NETZSCH, Selb, Germany) Simultaneous thermal analyzer (STA 429 CD) in a high purity argon flow at a gas flow rate of 50 mL per minute [30]. For the analysis, a salt mixture 35KCl–35NaCl–30$CaCl_2$–NdF_3, granular zinc and magnesium were weighed into an alundum melting pot. Two heating–cooling cycles were carried out at a rate of 10 °C per minute. In the first cycle, the melting pot with the weighed portion was heated to a temperature of 800 °C, then cooled to 200 °C; in the second successive cycle, the melting pot was heated from 200 to 800 °C and cooled again.

Experimental studies were carried out on a laboratory setup, which includes the following items: the shaft electric furnace with silicon carbide heating elements, the thermostat, the furnace control panel, the thermocouple and the stirring device. The synthesis of ternary master alloy Mg–Zn–Nd was carried out according to the following procedure. A mixture of salts consisting of Potassium chloride, (35 wt.%), Sodium chloride (35 wt.%), Calcium chloride (30 wt.%) was previously prepared. To this mixture was added neodymium fluoride (NdF_3) and the mixture stirred continuously. Then the resulting technological salt mixture was added to magnesium and zinc and placed in an alundum melting pot in the furnace. In the first series of experiments, in order to determine the optimal temperature condition for the synthesis of the master alloy, the melting pot was kept in a furnace with temperatures variations from 550 to 700 °C, and the residence time from 15 to 30 min, with a constant ratio of chlorides to neodymium fluoride. In the second series of experiments, in order to identify the most optimal technological condition, the synthesis temperature was varied from 650 to 700 °C, the residence time from 20 to 30 min, the ratio of chlorides to neodymium fluoride from 1:4 to 1:6. After the exposure time, in all experiments, the melt was settled for 5 min. The surface part of the molten salt was poured out, and the resulting master alloy was poured into molds. The studies were carried out with a constant mass ratio of Mg: Zn 1:2 and components of the salt mixture 35KCl–35NaCl–30$CaCl_2$, the qualifications of the initial salts of the mixture are: KCl, NaCl, $CaCl_2$,—chemically pure, NdF_3—pure. The average values of the degree extraction for neodymium obtained from the results of three parallel experiments are presented in the discussion of the results.

Elemental analysis of master alloy samples, which were obtained after DTA and experiments on a laboratory setup, was carried out on the sequential Wavelength Dispersive X-Ray Fluorescence Spectrometer XRF-1800 (Shimadzu, Kyoto, Japan). The metallographic study for the samples of the obtained master alloy was carried out on the Axiovert 40 MAT optical microscope (Carl Zeiss, Oberkochen, Germany) and the VEGA electron microscope (TESCAN, Brno, Czech Republic) with the INCAx-act energy dispersive spectrometer (Oxford Instruments, Abingdon, UK). The research was carried out with the involvement of the laboratory facilities of the Common Use Center of the Saint-Petersburg Mining University.

3. Results

At the first stage of experimental studies, DTA of the reduction process for neodymium from a technological salt mixture with a magnesium–zinc alloy was carried out. Figure 1 shows the thermograms obtained in the first heating cycle (green line of the curve) and the second heating cycle (purple line of the curve) of the Mg–Zn–KCl–NaCl–$CaCl_2$–NdF_3 sample up to 800 °C.

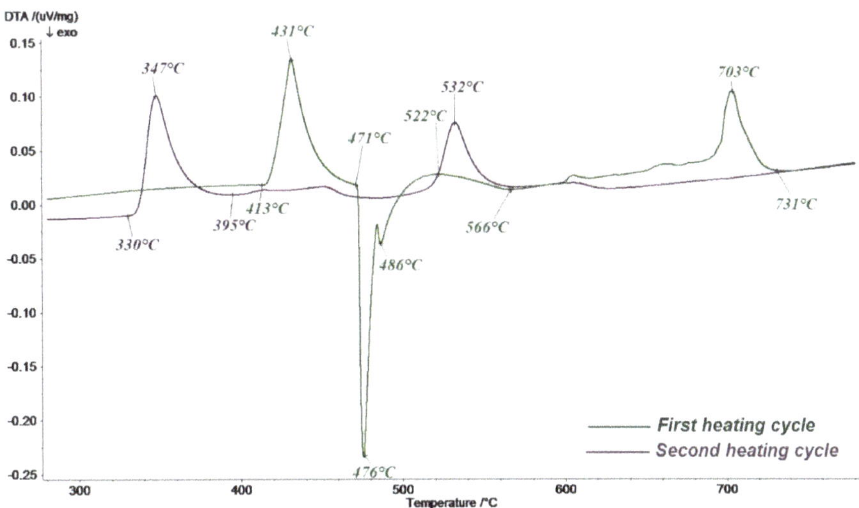

Figure 1. Thermograms of the first (green) and second (purple) heating cycle of the sample Mg–Zn–KCl–NaCl–$CaCl_2$–NdF_3 up to 800 °C.

During the first cycle of heating the charge, the beginning of zinc melting was recorded at a temperature of 413 °C. It was accompanied by an endothermic effect with a maximum at 431 °C. The interaction with magnesium begins after the zinc melts, which is confirmed by a strong exothermic effect with a minimum at 476 °C. It is typical for the process of formation of intermetallic compounds from pure elements. The exothermic peak with a minimum at 486 °C cannot be interpreted since it is associated with the uneven melting of the charge and the peculiarity of filling the melting pot. The onset of the reduction of neodymium by magnesium–zinc melt from fluoride in the chloride melt was recorded at a temperature of 522 °C after the end of the interaction of magnesium and zinc. This is confirmed by an extended exothermic effect with a minimum at 566 °C. The endothermic effect with a maximum at 703 °C corresponds to the melting of the components of the technological salt mixture, and at the temperature of 731 °C all interactions in the system under study cease. During the second heating cycle, the thermogram recorded endothermic effects of melting of the magnesium-zinc alloy with a maximum at 347 °C and a ternary master alloy with a maximum at 532 °C It was found that this effect corresponds to the melting of a ternary compound $Mg_xNd_yZn_z$ [7].

Figure 2 shows the thermograms obtained in the first cooling cycle (blue line) and the second cooling cycle (brown line) of the Mg–Zn–KCl–NaCl–$CaCl_2$–NdF_3 sample up to 200 °C.

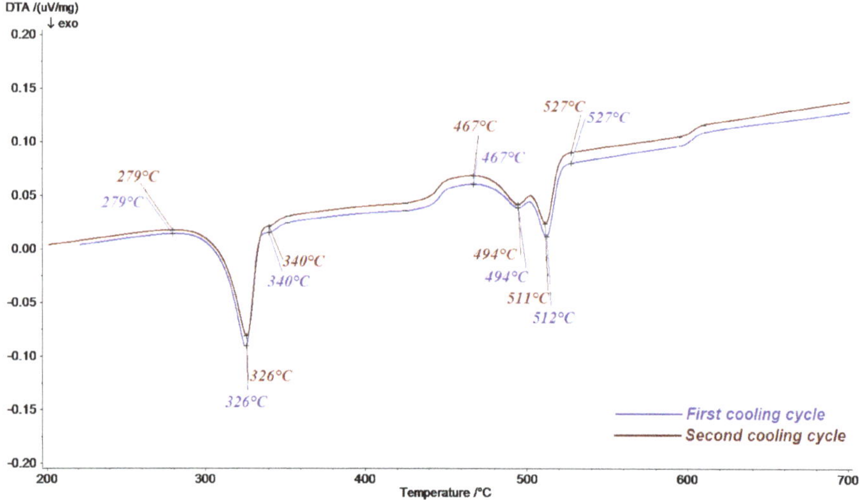

Figure 2. Thermograms of the first (blue) and second (brown) cooling cycle of the sample Mg–Zn–KCl–NaCl–$CaCl_2$–NdF_3 up to 200 °C.

During the first and second cooling cycles, two thermal effects were revealed on the thermograms. The first with the minimum at 326 °C, corresponding to the crystallization of the double magnesium–zinc eutectic. The second at 511–512 °C, corresponding to the crystallization of the $Mg_xNd_yZn_z$ ternary compound. In addition, thermal effects were recorded at the temperature of 494 °C, corresponding to the crystallization of the reacted technological salt mixture. Elemental analysis of the obtained sample for the magnesium–zinc–neodymium master alloy showed the presence of 18.11 wt.% neodymium, which proves the fundamental possibility of obtaining the master alloy at temperatures up to 700 °C and indicates the process of the reduction of neodymium from the fluoride–chloride melt with the magnesium–zinc alloy (Table 1).

Table 1. Chemical composition of Mg–Zn–Nd master alloy.

Master Alloy	Mass Fraction, %							
	Main Components			Impurities				
	Zn	Mg	Nd	Fe	Cu	Ni	Si	Al
Mg–Zn–Nd,	53.46	28.15	18.11	0.12	0.06	0.01	0.05	0.04

In the first series of experiments, the temperature effect on the degree of neodymium extraction was studied during experimental research on the neodymium reduction from a fluoride–chloride melt in a shaft electric furnace. In this case, in order to reduce the temperature for the synthesis of the ternary master alloy was taken the constant value of the Mg:Zn ratio equal to 1:2 [31]. As a result of processing the obtained data, the dependences of the degree of extraction of neodymium on the residence time at temperatures of 550, 600, 650, 700 °C were assessed (Figure 3).

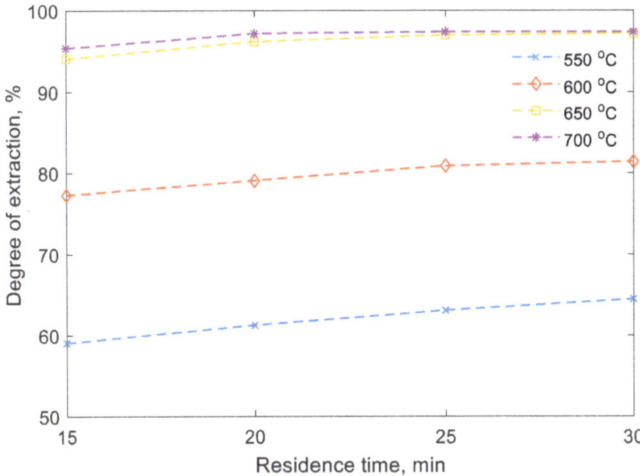

Figure 3. Dependence of the degree of extraction of neodymium and the residence time at temperatures of 550–700 °C.

During the experiments, it was confirmed that the addition of zinc to the charge helps to reduce the temperature of the neodymium reduction process from the molten salt, in contrast to the temperature ranges for obtaining the double magnesium–neodymium master alloy [32]. It was found that the degree of extraction of neodymium of up to 60% is achieved at a synthesis temperature of 550 °C. Moreover, the degree of extraction of neodymium increases to 92.2–93.2% with an increase in temperature to 650 °C

In the second series of experiments, studies were carried out in order to identify the most optimal technological parameters for conducting melts. In this case, a high degree of neodymium extraction into the maser alloy is achieved. The initial data and the results of experiments on obtaining the master alloy are shown in Table 2.

Table 2. Results of synthesis of Mg–Zn–Nd master alloy.

Melt Number	Chloride Ratio: NdF_3	T, °C	t, min	Stirring	Nd Recovery, %
1	4:1	650	20	no	86.9
2	6:1	650	20	no	95.8
3	4:1	700	20	yes	88.1
4	6:1	700	20	yes	99.6
5	4:1	650	30	yes	88.4
6	6:1	650	30	yes	99.5
7	4:1	700	30	no	88.1
8	6:1	700	30	no	96.3
9	4:1	650	20	yes	86.2
10	6:1	650	20	yes	99.6
11	4:1	700	20	no	85.2
12	6:1	700	20	no	91.4
13	4:1	650	30	no	84.3
14	6:1	650	30	no	93.6
15	4:1	700	30	yes	87.9
16	6:1	700	30	yes	99.6

It was determined that the neodymium fluoride reduction is accompanied by the formation of the homogeneous magnesium–zinc–neodymium master alloy. According to experimental data, it has been proved that the neodymium yield increases to 99.6% with an increase in the ratio of chlorides to NdF_3

up to 6:1 in the technological salt mixture and with continuous stirring of the melt. With an increase in temperature up to 700 °C, the neodymium yield does not change significantly. As a result of the performed melts, master alloys with a neodymium content of 10 to 25 wt.% were obtained. Analysis of the quality of the resulting master alloy showed that its macrostructure is characterized by the absence of gas pores and non-metallic inclusions (Figure 4).

Figure 4. Macrostructure of master alloy 25Mg–50Zn–25Nd.

Since there are no regulatory requirements for the ternary master alloy of the studied composition, the comparison was made with the requirements for the content of impurities for the magnesium–neodymium alloy; this one is also used for the production of special-purpose magnesium alloys. The Mg–Zn–Nd master alloys meets the requirements for Mg–Nd magnesium master alloys according to specification TU 48-4-271-91 (Table 3).

Table 3. Chemical composition of Mg–Zn–Nd master alloy.

Alloy Grade	Mass Traction, %							
	Main Components			Impurities, max				
	Zn	Mg	REEs	Fe	Cu	Ni	Si	Al
MH	-	base	20–35	0.15	0.1	0.01	0.05	0.05
Mg–Zn–Nd, melting №16	base	25.0	24.8	0.08	0.06	-	0.03	0.02

The microstructure of all tested master alloys in the cast state consisted of a magnesium–zinc matrix, and well-distinguishable individual intermetallic compounds, mainly polygon-shaped. Regular shapes and some coagulation of the edges can be detected, which ultimately leads to the absence of local defects in the microstructure in the matrix. A significant part of the thin section for the obtained magnesium–zinc–neodymium master alloy with a content of 24.8 wt.% neodymium (Figure 5a) is occupied by eutectic colonies (dark areas). It is located along the boundaries of the individual intermetallic compounds (light areas). In some cases, areas of accretion of intermetallic compounds are observed. The average grain diameter is 35 μm. Zones of eutectics are revealed at ×500 magnification (Figure 5b). They are located within the individual intermetallic compounds.

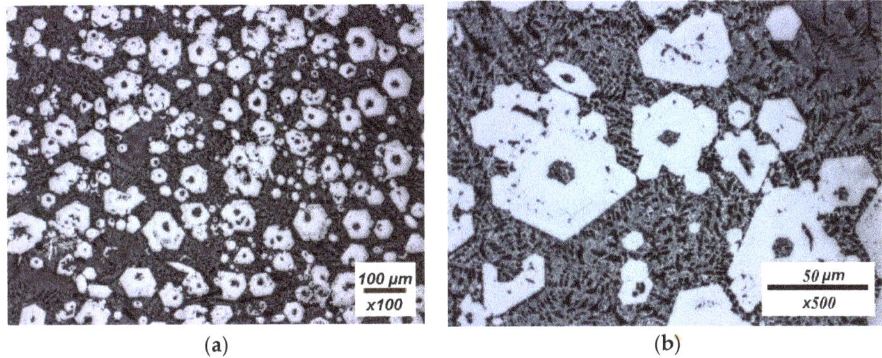

(a) (b)

Figure 5. Microstructure of master alloy 25Mg–50Zn–25Nd. Zoom (**a**) ×100, and (**b**) ×500.

Micro X-ray spectral analysis of the structure sections shows (Figure 6a) that individual intermetallic compounds contain about 21 wt.% neodymium, which corresponds to the $Mg_{3,4}NdZn_7$ phase. Double eutectic (Figure 6b), alloyed with neodymium, contains 32.41 wt.% magnesium, 60.72 wt.% zinc and 6.88 wt.% neodymium (Table 4).

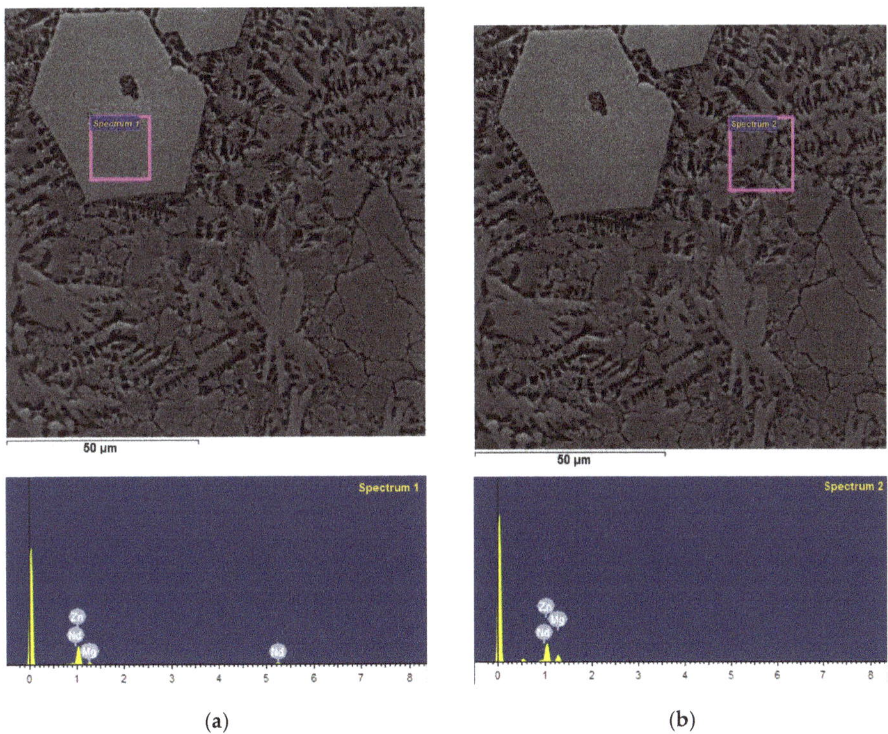

(a) (b)

Figure 6. Electronic images of the microstructure of the 25Mg–50Zn–25Nd master alloy (**a**) and (**b**)—×2000.

Table 4. Values of spectra of Mg–Zn–Nd master alloy.

Spectrum	Mg, wt.%.	Zn, wt.%.	Nd, wt.%.
spectrum 1 (a)	12.08	66.73	21.19
spectrum 2 (b)	32.41	60.72	6.88

4. Conclusions

Thus, as a result of the experimental studies, a ternary master alloy of magnesium–zinc–neodymium was obtained, while optimal technological conditions of the synthesis were revealed (temperature 650 °C, residence time 20 min, ratio of chlorides to neodymium fluoride 1:6, while stirring the melt), in this case, the degree of extraction of neodymium achieved is up to 99.6%.

It was established that the onset of the reduction of neodymium by magnesium–zinc melt from its fluoride in the chloride melt occurs at a temperature of 522 °C using DTA. This is confirmed by an extended exothermic effect with a minimum at 566 °C.

Analysis of the microstructure showed that the obtained ternary master alloy 25Mg–50Zn–25Nd is characterized by a uniform distribution of intermetallic compounds (Mg_3NdZn_6) in the bulk of the double magnesium–zinc eutectic. The obtained experimental data are a prerequisite for the development of industrial technology for the production of magnesium–zinc–neodymium alloys for their use in non-ferrous and ferrous metallurgy.

Author Contributions: Conceptualization, I.B.; methodology, S.S.; software, I.B.; validation, S.S., I.B., R.K.; formal analysis, V.B.; investigation, I B.; resources, S.S.; writing—original draft preparation, S.S.; writing—review and editing, R.K. and V.B.; visualization, I.B.; supervision, R.K.; project administration, I.B.; funding acquisition, I.B. All authors have read and agreed to the published version of the manuscript.

Funding: This research received no external funding.

Acknowledgments: The authors thank St. Petersburg Mining University for the opportunity to realize research. The studies were performed using the equipment of the Common Use Centre of the Saint-Petersburg Mining University.

Conflicts of Interest: The authors declare no conflict of interest.

References

1. Solomon, E.L.S.; Araullo-Peters, V.; Allison, J.E.; Marquis, E.A. Early precipitate morphologies in Mg-Nd-(Zr) alloys. *Scr. Mater.* **2017**, *128*, 14–17. [CrossRef]
2. Chen, J.K.; Chen, Y.C.; Hsien-Tsung, L.I.; Chan, K.S.; Chang, C.J. Effects of Nd and rotary forging on mechanical properties of AZ71 Mg alloys. *Trans. Nonferrous. Met. Soc. China* **2015**, *25*, 3223–3231. [CrossRef]
3. Zhou, B.; Wang, L.; Chen, B.; Jia, Y.; Wen, W.; Li, D.; Shu, D.; Jin, P.; Zeng, X.; Ding, W. Study of age hardening in a Mg–2.2 Nd alloy by in situ synchrotron X-ray diffraction and mechanical tests. *Mater. Sci. Eng.* **2017**, *708*, 319–328. [CrossRef]
4. Jin, S.; Zhang, D.; Lu, X.; Zhang, Y.; Tan, L.; Liu, Y.; Wang, Q. Mechanical properties, biodegradability and cytocompatibility of biodegradable Mg-Zn-Zr-Nd/Y alloys. *J. Mater. Sci. Technol.* **2020**, *47*, 190–201. [CrossRef]
5. Wang, W.; Wu, D.; Chen, R.; Qi, Y.; Ye, H.; Yang, Z.Q. Revisiting the role of Zr micro-alloying in a Mg-Nd-Zn alloy. *J. Alloy Compd.* **2020**, *832*, 155016. [CrossRef]
6. Sheng, L.; Du, B.; Hu, Z.; Qiao, Y.; Xiao, Z.; Wang, B.; Xu, D.; Zheng, Y.; Xi, T. Effects of annealing treatment on microstructure and tensile behavior of the Mg-Zn-Y-Nd alloy. *J. Magnes. Alloy* **2020**, *8*, 601–613. [CrossRef]
7. Zhu, Z.; Gharghouri, M.A.; Pelton, A.D. Thermodynamic modelling and in-situ neutron diffraction investigation of the (Nd+Mg+Zn) system. *J. Chem. Thermodyn.* **2016**, *94*, 43–51. [CrossRef]
8. Li, Z.; Liu, F.; Yuan, A.; Duan, B.; Li, Y.; Li, X. Effect of rolling deformation on microstructure and texture of spray-deposited magnesium alloy containing Mg-Nd-Zn typed LPSO. *J. Mater. Sci. Technol.* **2017**, *33*, 630–636. [CrossRef]

9. Peng, Y.-G.; Du, Z.-W.; Liu, W.; Li, Y.-J.; Li, T.; Han, X.-L.; Ma, M.-L.; Pang, Z.; Yuan, J.-W.; Shi, G.-L. Evolution of precipitates in Mg−7Gd−3Y−1Nd−1Zn−0.5Zr alloy with fine plate-like 14H-LPSO structures aged at 240 °C. *Trans. Nonferrous Met. Soc. China* **2020**, *30*, 1500–1510. [CrossRef]
10. Yu, Z.; Xu, C.; Meng, J.; Zhang, X.; Kamado, S. Microstructure evolution and mechanical properties of as-extruded Mg-Gd-Y-Zr alloy with Zn and Nd additions. *Mater. Sci. Eng. A* **2018**, *713*, 234–243. [CrossRef]
11. Lan, A.; Huo, L. Effect of substitution of minor Nd for Y on mechanical and damping properties of heat-treated Mg–Zn–Y–Zr alloy. *Mater. Sci. Eng. A* **2016**, *651*, 646–656. [CrossRef]
12. Guobing, W.; Xiaodong, P.; Junchen, L.; Weidong, X.; Qunyi, W. Structure Heredity Effect of Mg-10Y Master Alloy in AZ31 Magnesium Alloy. *Rare Met. Mater. Eng.* **2013**, *42*, 2009–2013. [CrossRef]
13. Khrustalyov, A.P.; Garkushin, G.V.; Zhukov, I.A.; Razorenov, S.V.; Vorozhtsov, A.B. Quasi-Static and Plate Impact Loading of Cast Magnesium Alloy ML5 Reinforced with Aluminum Nitride Nanoparticles. *Metals* **2019**, *9*, 715. [CrossRef]
14. Kaminsky, V.V.; Petrovich, S.A.; Lipin, V.A. Production of intermetallides in the Al–Ti–Zn system. *J. Min. Inst.* **2018**, *233*, 512–517.
15. Glavatskikh, M.V.; Pozdnyakov, A.V.; Makhov, S.V.; Napalkov, V.I. Investigation into the structure and phase composition of powder aluminum-phosphorus master alloys. *Russ. J. Nonferrous Met.* **2014**, *55*, 450–455. [CrossRef]
16. Xiaodong, P.; Junchen, L.; Sunyun, X.; Guobin, W.; Yan, Y. Effects of Different State Mg-5Sr-10Y Master Alloys on the Microstructure Refinement of AZ31 Magnesium Alloy. *Rare Met. Mater. Eng.* **2013**, *42*, 2421–2426. [CrossRef]
17. Savchenkov, S.A.; Bazhin, V.Y.; Povarov, V.G. Research on the process of gadolinium recovery from the melt of salts on formation of Mg–Zn–Gd master alloys for manufacturing of magnesium and aluminium special-purpose alloys. *Nonferrous Met.* **2020**, *48*, 35–40.
18. Zhang, Y.; Ye, C.; Shen, Y.; Chang, W.; StJohn, D.; Wang, G.; Zhai, Q. Grain refinement of hypoeutectic Al-7wt.%Si alloy induced by an Al–V–B master alloy. *J. Alloy Compd.* **2020**, *812*, 152022. [CrossRef]
19. Jiang, Z.; Meng, X.; Jiang, B.; Jiang, S.; Dai, J.; Dong, J.; Ding, Y. Grain refinement of Mg-3Y alloy using Mg-10Al2Y master alloy. *J. Rare Earths* **2020**, in press. [CrossRef]
20. Xiaoyan, W.; Zhang, H.; Feng, J.; Ying, Y.; Lina, J.; Hu, Z. Microstructure and Grain Refinement Performance of a New Al-5Nb-RE-B Master Alloy. *Rare Met. Mater. Eng.* **2018**, *47*, 2017–2022. [CrossRef]
21. Order of the Ministry of Industry and Trade of the Russian Federation of May 5, 2014 No. 839 On Approval of the Strategy for the Development of the Ferrous Metallurgy of Russia for 2014-2020 and for the Prospect until 2030 and the Strategy for the Development of the Non-Ferrous Metallurgy of Russia for 2014–2020 and for the Prospect until 2030. Available online: https://www.garant.ru/products/ipo/prime/doc/70595824/ (accessed on 10 September 2020).
22. Litvinenko, V.S. Digital Economy as a Factor in the Technological Development of the Mineral Sector. *Nat. Resour. Res.* **2019**, *28*, 1–21. [CrossRef]
23. Yurak, V.; Dushin, A.; Mochalova, L. Vs sustainable development: scenarios for the future. *J. Min. Inst.* **2020**, *242*, 242. [CrossRef]
24. El-Deeb, A.B.; Brichkin, V.N.; Kurtenkov, R.V.; Bormotov, I.S. Extraction of Alumina from Kaolin: A Comparative Study of Pyrometallurgical and Hydrometallurgical Processes. *Appl. Clay Sci.* **2019**, *172*, 146–154. [CrossRef]
25. Nevskaya, M.A.; Marinina, O.A. Challenges and opportunities of state regulation of the innovation process in the Russian mineral resources sector. *Acad. Strateg. Manag. J.* **2017**, *16*, 149–159.
26. Elbendary, A.; Aleksandrova, T.; Nikolaeva, N. Influence of operating parameters on the flotation of the Khibiny Apatite-Nepheline Deposits. *J. Mater. Res. Technol.* **2019**, *8*, 5080–5090. [CrossRef]
27. Nevskaya, M.A.; Seleznev, S.G.; Masloboev, V.A.; Klyuchnikova, E.; Makarov, D.V. Environmental and Business Challenges Presented by Mining and Mineral Processing Waste in the Russian Federation. *Minerals* **2019**, *9*, 445. [CrossRef]

28. Litvinenko, V.S.; Tsvetkov, P.S.; Molodtsov, K.V. The social and market mechanism of sustainable development of public companies in the mineral resource sector. *Eurasian Min.* **2020**, 36–41. [CrossRef]
29. Kondrasheva, N.K.; Rudko, V.A.; Nazarenko, M.Y.; Povarov, V.G.; Derkunskii, I.O.; Konoplin, R.R.; Gabdulkhakov, R.R. Influence of Parameters of Delayed Coking Process and Subsequent Calculation on the Properties and Morphology of Petroleum Needle Coke from Decant Oil Mixture of West Siberian Oil. *Energy Fuels* **2019**, *33*, 6373–6379. [CrossRef]
30. Aleksandrova, T.N.; Heide, G.; Afanasova, A.V.; Academy, F.M.; Saint-Petersburg Mining University. Assessment of refractory gold-bearing ores based of interpretation of thermal analysis data. *J. Min. Inst.* **2019**, *235*, 30–37. [CrossRef]
31. Savchenkov, S.; Bazhin, V.; Brichkin, V.; Povarov, V.; Ugolkov, V.; Kasymova, D. Synthesis of magnesium-zinc-yttrium master alloy. *Lett. Mater.* **2019**, *9*, 339–343. [CrossRef]
32. Savchenkov, S.A.; Bazhin, V.Y.; Brichkin, V.N.; Kosov, Y.I.; Ugolkov, V.L. Production Features of Magnesium-Neodymium Master Alloy Synthesis. *Mettalurgist* **2019**, *63*, 394–402. [CrossRef]

Article

Thermal Expansion of $MgTiO_3$ Made by Sol-Gel Technique at Temperature Range 25–890 °C

Tamir Tuval [1], Brian A. Rosen [2], Jacob Zabicky [3], Giora Kimmel [1,3], Helena Dilman [1] and Roni Z. Shneck [1,*]

1 Department of Materials Engineering, Ben-Gurion University of the Negev, Beer-Sheva 84105, Israel; tuvalt@gmail.com (T.T.); gyorakimmel@gmail.com (G.K.); hdilman@bgu.ac.il (H.D.)
2 Department of Materials Science and Engineering, Tel Aviv University, Ramat Aviv 69978001, Israel; barosen@post.tau.ac.il
3 Institutes for Applied Research, Ben-Gurion University of the Negev, Beer-Sheva 84105, Israel; zabicky@exchange.bgu.ac.il
* Correspondence: roni@bgu.ac.il

Received: 2 September 2020; Accepted: 29 September 2020; Published: 1 October 2020

Abstract: $MgTiO_3$ is a material commonly used in the industry as capacitors and resistors. The high-temperature structure of $MgTiO_3$ has been reported only for materials synthesized by the solid-state method. This study deals with $MgTiO_3$ formed at low temperatures by the sol-gel synthesis technique. Co-precipitated xerogel precursors of nanocrystalline magnesium titanates, with Mg:Ti ratio near 1:1, were subjected to thermal treatment at 1200 °C for 5 h in air. A sample with fine powders of $MgTiO_3$ (geikielite) as a major phase with Mg_2TiO_4 (qandilite) as a minor phase was obtained. The powder was scanned on a hot-stage X-ray powder diffractometer at temperatures between 25 and 890 °C. The lattice parameters and the atomic positions of the two phases were determined as a function of temperature. The thermal expansion coefficients of the geikielite were derived and compared with previously published data using the solid-state synthesis technique, providing insights on trends in materials properties at elevated temperature as a function of synthesis. It was found that the deviation of the present results in comparison to previously reported data do not originate from the method of synthesis but rather from the fact that there is an asymmetric solubility gap in geikielite. The lattice parameters of this study present the property of stoichiometric $MgTiO_3$ and are compared to previously reported non-stoichiometric $MgTiO_3$ with excess of Ti. The values of lattice parameters of the non-stoichiometric versus temperature of geikielite found the same for both solid-state reaction and sol-gel products.

Keywords: $MgTiO_3$; geikielite; high-temperature X-ray diffraction; sol-gel technique; thermal expansion

1. Introduction

The MgO-TiO_2 ceramic system includes three stoichiometric magnesium titanate phases: Mg_2TiO_4 (qandilite), $MgTiO_3$ (geikielite) and $MgTi_2O_5$ (karrooite) [1]. Their crystal structures have been determined by X-ray diffraction and neutron diffraction [2,3]. The phases are:

(a) Geikielite, $MgTiO_3$, is rhombohedral of ilmenite type, space group R-3 (148), Pearson Symbol hR10.0; formed at temperatures above 600 °C, is stable from room temperature to its melting point;
(b) Qandilite, Mg_2TiO_4, cubic of inverse spinel type, space group Fd-3m (227), Pearson Symbol cF56.0, formed at above 1150 °C;
(c) Karrooite, $MgTi_2O_5$, orthorhombic of pseudo-brookite type, space group Cmcm (63), Pearson Symbol oC32.0 formed above 500 °C.

High-temperature X-ray diffraction (HT-XRD) studies have been published for all three stoichiometric Mg titanates prepared by the sol-gel technique during their transformation from the xerogel precursor to their final oxide form for a range of temperatures between 700 and 1300 °C [4]. High-temperature neutron diffraction (HT-ND) studies have been made on Mg_2TiO_4 synthesized by the solid-state reaction at temperatures between 90 to 1400 °C [5]. In this study, there was agreement between the lattice parameter of the HT-XRD and HT-ND data in the temperature range between 800 and 1300 °C [6]. Importantly, the in-situ HT-XRD studies of $MgTiO_3$ during the firing of xerogel powders between 700 and 1300 °C yielded linear thermal expansion coefficients (TECs).

There are two additional publications with ND-XRD data for $MgTiO_3$, which by contrast, showed non-linear TECs [7,8]. Both publications are for the same sample -Kar2-, studied where the $MgTi_2O_5$ (karrooite) phase was the major component and the $MgTiO_3$ (geikielite) and TiO_2 (rutile) phases were the minor components (by mass). In these publications, the phases were obtained by the solid-state reaction between $MgCO_3$ and TiO_2 fine powders, and the HT-ND data [7,8] were collected on sample Kar2 between 23 and 1305 °C.

Further examination of the HT diffraction works on geikielite showed that there is a significant gap in the lattice parameters between the HT-XRD data received from sol-gel product [4], and HT-ND data received from geikielite using the solid state reaction in that the lattice parameters of the sol-gel product were slightly lower. A key difference in the collection of the data was that the sol-gel data were obtained using a different sample at each temperature after 1 h firing, and the solid-state version was a single sample for all temperatures. In order to determine if the gap originated from the experimental methods or inherent differences in the synthesis (sol-gel vs. solid state), it was decided to investigate a single geikielite sample determining lattice parameters between RT up to 900 °C.

In this work, we complete the HT crystallographic data of sol-gel product of $MgTiO_3$ using HT-XRD measurements of $MgTiO_3$ made by sol-gel technique between 25 and 900 °C and provide insight to the differences in TEC behavior.

2. Materials and Methods

2.1. Synthesis

The magnesium titanates were prepared by the sol-gel method using metalorganic precursors: diethyl ethoxymagnesiomalonate, prepared by metallation of diethyl malonate (1), and titanium(IV) tetra-tert-butoxide (Aldrich), dissolved in anhydrous 2-propanol. The hydrolysis step was carried out at nearly room temperature, using a stream of hot air (about 100 °C) containing water vapor (superheated steam), during about 3 hours, to ensure total hydrolysis. The precipitated solid was filtered, washed with 2-propanol and left to dry in air for several days, until a constant weight was obtained. After analysis of the Mg and Ti content, this solid served as starting material for thermal treatment.

The powders were initially fired at 600 °C for 3 h. After the initial treatment, the sample with Mg:Ti 1.1:1 found as a single geikielite phase, where a = 5.0537 (3) Å; c = 13.897 (5) Å; atomic positions: z(Mg) = 0.3563 (2); z(Ti) = 0.1445 (2); x(O) = 0.315(1); y(O) = 0.0225(1); z(O) = 0.247(1). However, the expected Mg:Ti ratio was confirmed by inductively coupled plasma (ICP). The amount of impurities was below 10 ppm. Additional treatments were made at 1200 °C for 5 h in order to obtain well-crystallized powders. After the 1200 °C treatment, the sample was found with 95% geikielite and 5% qandilite. It was designated as "GQ" (major phase geikielite and minor phase qandilite) and will be referred to as such throughout the manuscript. The qandilite within GQ served as the internal standard.

2.2. RT XRD Measurements

The samples for HT-XRD studies were characterized by a Rigaku powder X-Ray Diffractometer. Data were collected in the conventional Bragg–Brentano configuration (theta/2theta) by means of Cu K_α radiation at 40 kV and 30 mA. The K_β was filtered out by graphite monochromator attached to the detector. Phase characterization from XRD data was made by using public domain FullProf/WinPlotter software [9]).

2.3. HT XRD Measurements

X-ray diffraction was performed on a Bruker D8 Advance in Bragg–Brentano geometry using an X-ray source with a Cu anode having a $K_{\alpha 1}$ emission wavelength of 1.5406 angstroms. Samples were placed in an Anton Paar XRK-900 high-temperature reaction chamber using a Macor sample stage. The influence of the thermal expansion of the stage was measured by calibrating the stage height as a function of temperature using an alignment slit. The temperature of the sample was controlled using and Anton Paar TCU 750 controller by mounting a K-type thermocouple in the sample holder adjacent to the sample. A linear PSD detector (LYNXEYE XE-T) was used with an opening of 2.94 degrees. The diffraction pattern was recorded from two-theta of 10 degrees until 120 degrees using a coupled theta/two-theta scan type. Data points were acquired in increments of 0.02 degrees with an acquisition time of 0.25 s. Lattice parameters were fitted using TOPAS software with a TCHZ function. The refined parameters included the lattice parameters, sample displacement and zero error. The line position and effect of instrumental broadening and asymmetry were calibrated by SRM 660c LaB_6.

The diffractograms were also analyzed by the program Powder-Cell [10]. In this step, the phases were easily identified, and the unit cells were verified for the thermal expansion. Grain shape and size was assessed by HR-SEM.

2.4. Thermal Expansion Methodology

It is essential to know the dimensions of ceramic materials as function of temperatures in order to calculate the dimensions of objects working at elevated temperatures and to evaluate thermal stresses during temperature changes. For practical reasons it is suggested to define the relative dimension change of a material by Equation (1):

$$[L(T) - L_0]/L_0 = f(T - T_0) \quad (1)$$

where $L(T)$ is the size of one dimension at temperature T, L_0 is the size of the same dimension at ambient temperature, T is the working temperature and T_0 is the ambient temperature (usually the ambient temperature is 25 °C).

In the case of linear thermal expansion, Equation (1) becomes

$$[L(T) - L_0]/L_0 = \alpha(T - T_0) \quad (2)$$

where α is the linear thermal expansion coefficient. For a general case it is possible to define

$$\Delta L/L_0 = \alpha_1\, \Delta T + \alpha_2\, \Delta T^2 + \cdot\cdot + \alpha_n\, \Delta T^n \quad (3)$$

or

$$L(T) = L_0(1 + \alpha_1\, \Delta T + \alpha_2\, \Delta T^2 + \cdot\cdot + \alpha_n\, \Delta T^n) \quad (4)$$

where

$$\Delta L = L(T) - L_0 \text{ and } \Delta T = T - T_0$$

Neglecting the contribution of enhanced vacancy formation at higher temperatures to the size of a sample of matter, the thermal expansion of a periodical crystalline matter can be modeled by measuring its lattice parameters as function of temperature. By using HT diffraction, each lattice parameter, A^j, can be obtained directly from the measurement as

$$A^j(T) = A^j{}_0 + k_1\, \Delta T + k_2\, \Delta T^2 + \cdot\cdot + k_n\, \Delta T^n \quad (5)$$

In this case, the non-linear thermal coefficients will be given as

$$\alpha_i^j = \mathrm{k_i}/A^j{}_0 \tag{6}$$

From Equation (3) it is then possible to define an overall thermal expansion coefficient along a crystal axis A^j as

$$\alpha_{A^j} = \Delta A^j/(A^j{}_0\ \Delta T) = \alpha_1 + \alpha_2\ \Delta T + \cdots + \alpha_n\ \Delta T^{n-1} \tag{7}$$

Similarly, for unit cell volume V we define the volumetric thermal expansion as

$$\gamma = \Delta V/(V_0\ \Delta T) = \gamma_1 + \gamma_2\ \Delta T + \cdots + \gamma_n\ \Delta T^{n-1} \tag{8}$$

It should be noted that this methodology is valid only where there is no phase transformation or significant crystal structure change in the range of measurements.

3. Results

3.1. As-Received Sample

HR-SEM showed that GQ sample contained fine powder with average grain size at the range of 250–350 nm. This yielded high-quality XRPD diffractograms without preferred orientation

The crystal structures of Mg_2TiO_4 and $MgTiO_3$ in sample GQ after 5 h annealing at 1200 °C and scanned at room temperature by XRD and analyzed by Rietveld software are given below, in comparison with previously published data. The Rietveld diagram is shown in Figure 1. The phase amounts in sample GQ found as 95.5 wt% $MgTiO_3$ with 4.5 wt% $MgTiO_4$.

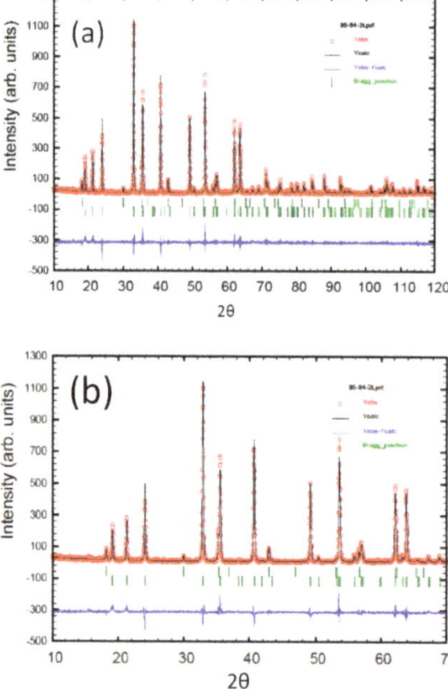

Figure 1. (**a**) Rietveld diagram (FullProf) of sample GQ after 5 h annealing at 1200 °C and scanned in a Rigaku diffractometer at room temperature (whole 2θ range). (**b**) Rietveld diagram (FullProf) of sample GQ after 5 h annealing at 1200 °C and scanned in Rigaku diffractometer at room temperature (small 2θ range).

The lattice parameters of $MgTiO_3$ in sample GQ at room temperature after annealing at 1200 °C for 5 h in comparison with ICDD data base are given in Table 1.

Table 1. RT lattice parameter of $MgTiO_3$.

Lattice Parameters		Lattice Volume	ICDD #	Comment
a [Å]	**c [Å]**	**V [Å³]**		**Quality**
5.057	13.903	307.9	01-073-7748	Star
5.057	13.903	307.9	01-075-8341	Star
5.054	13.899	307.5	01-79-0831	Star
5.056 (2)	13.902 (2)	307.8 (3)	ICDD	Average
5.054 (1)	13.904 (2)	307.6 (3)	GQ AR	Reitveld, FullProf
5.054 (1)	13.902 (2)	307.5 (3)	GQ AR	Reitveld, Topas *

* Rwp 11.88, χ^2 = 1.34.

The lattice parameters of Mg_2TiO_4 in sample GQ at room temperature after annealing at 1200 °C for 5 h in comparison with ICDD data base are given in Table 2.

Table 2. Room temperature lattice parameters of Mg_2TiO_4.

Lattice Parameter	ICDD #	Comment
a [Å]		**Quality**
8.442	01-072-6966	Star
8.447	01-072-6977	Star
8.447	01-72-6967	Star
8.441	00-025-1157	Star
8.444 (3)	ICDD	Average
8.442 (1)	GQ AR	Ambient
8.441 (1)	GQ AR	Hot stage *

* Rwp 11.88, χ^2 = 1.34.

The atomic positions of $MgTiO_3$ in sample GQ at room temperature as refined after annealing at 1200 °C for 5 h in comparison with Reference [3] are given in Table 3 (*Rwp 11.88, χ^2 = 1.34).

Table 3. Ion positions in $MgTiO_3$ as received. (*) Hot stage before heat treatment; (**) Rigaku conventional diffractometer.

		$MgTiO_3$	
ion/position	GQ- *	GQ- **	Reference [3]
z(Mg^{2+})	0.3561	0.3556	0.3556
z(Ti^{4+})	0.1445	0.1446	0.1451
x(O^{2-})	0.3108	0.3173	0.3159
y(O^{2-})	0.0140	0.0215	0.0215
z(O^{2-})	0.2491	0.2465	0.2470

The atomic position of Mg_2TiO_4 in sample GQ at room temperature after annealing at 1200 °C for 5 h in comparison with Reference [3] is given in Table 4.

Table 4. The atomic position of Mg_2TiO_4 (XRD data from Rigaku conventional diffractometer).

	Mg_2TiO_4	
Ion/position	GQ	Reference [3]
z(O^{2-})	0.2588	0.2605

3.2. High-Temperature XRD (HT-XRD)

The lattice parameters versus temperature for the sample GQ containing geikielite and qandilite are listed in Table 5. The lattice parameters for the sample, which was scanned after 1 h in previous study [4], are given in Table 6.

Table 5. Lattice parameters of $MgTiO_3$ and Mg_2TiO_4 as function of temperature in the present investigation (sample GQ). All parameters are given in Å.

	$MgTiO_3$		Mg_2TiO_4
T °C	a * [Å]	c # [Å]	a # [Å]
25	5.054	13.902	8.441
50	5.055	13.905	8.443
100	5.056	13.913	8.445
200	5.061	13.929	8.453
400	5.071	13.963	8.471
500	5.076	13.980	8.480
550	5.079	13.989	8.484
600	5.081	13.998	8.489
650	5.084	14.007	8.494
700	5.087	14.016	8.499
800	5.092	14.035	8.510
890	5.098	14.054	8.520

* uncertainty = 0.0003 Å. # uncertainty = 0.0008 Å.

Table 6. Lattice parameters of $MgTiO_3$ as function of temperature in a previous investigation [4] of sol-gel product.

	Geikielite	
T °C	a [Å]	c [Å]
700	5.087	14.014
800	5.092	14.035
900	5.098	14.050
1000	5.104	14.072
1100	5.111	14.093
1200	5.117	14.115
1300	5.123	14.136

A uncertainty = 0.0002 Å. c uncertainty = 0.0005 Å.

Figure 2 shows that the lattice parameters versus temperature of the Mg_2TiO_4 perfectly integrated with literature data obtained by HT-ND [5].

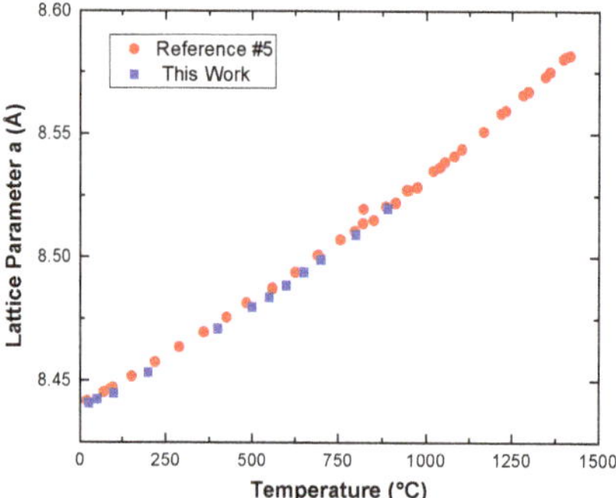

Figure 2. Lattice parameter a of the reference Mg_2TiO_4 material versus temperature from HT-XRD of sample GQ in comparison to reported HT-ND data [5].

3.3. Thermal Expansion Coefficients for the Reference Material (Mg_2TiO_4)

The lattice parameters versus T-25 (°C) were fitted as a polynomial $A^j(T) = A^j_0 + k_1 (T - 25) + k_2 (T - 25)^2$ as in Equation (5). Then, using Equation (6), $\alpha_i^j = k_i/A^j_0$ obtaining $\alpha_A{}^j = (A^j\text{-} A^j_{0)}/[A^j_0 (T - 25)] = \alpha_1 + \alpha_2 (T - 25)$ as given in Equation (7), or $\gamma = (V^j\text{-} V^j_{0)}/[V^j_0 (T - 25)] = \gamma_1 + \gamma_2 (T - 25)$ as given in Equation (8). The thermal expansion coefficients (TEC) for the Mg_2TiO_4 were calculated from the data of Table 1 and compared with published data [5] made by ND with the selection of similar temperature range. Equation (9) shows the TECs of Mg_2TiO_4 from this study HT-XRD (GQ sample) 25–890 °C

$$\begin{aligned} \alpha_a &= 8.3 \times 10^{-6} + 2.9 \times 10^{-9} (T - 25) \; [1/°C] \\ \gamma &= 2.5 \times 10^{-5} + 9.3 \times 10^{-9} (T - 25) \; [1/°C] \end{aligned} \tag{9}$$

Since the uncertainty of the lattice parameter is ~0.1% and the uncertainty in the temperature is ~0.2%, the maximum uncertainty of the terms in the reported equations for thermal expansion coefficient is ~0.3%. Equation (10) shows the TEC of Mg_2TiO_4 from previously published data [5], studied by HT-ND for selected temperature range 25–890 °C.

$$\begin{aligned} \alpha_a &= 9.2 \times 10^{-6} + 1.9 \times 10^{-9} (T - 25) \; [1/°C] \\ \gamma &= 2.7 \times 10^{-5} + 6.2 \times 10^{-9} (T - 25) \; [1/°C] \end{aligned} \tag{10}$$

3.4. Thermal Expansion Coefficients for $MgTiO_3$

There was excellent agreement between the high-temperature lattice parameter data of $MgTiO_3$ made in the present investigation by comparison to the results found in the higher temperature range for xerogels from the HT-XRD in situ study of seven samples [4], as shown in Table 6. It seems that the presence of qandilite did not affect the TECs of the geikielite. HT-XRD data of the GQ sample together with published data [4,8] of $MgTiO_3$ are plotted in Figure 3.

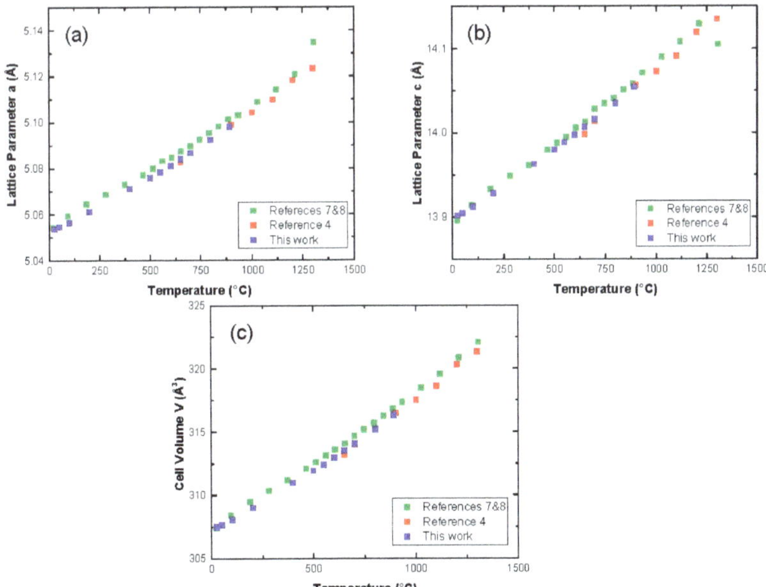

Figure 3. (**a**) Lattice parameter a, (**b**) lattice parameter c and (**c**) cell volume versus temperature for $MgTiO_3$ from present GQ and previous samples HT diffraction. Data taken from this work and [4,7,8].

For the present HT-XRD study, with the data collected at a temperature range between 25 and 890 °C, the overall thermal expansion coefficients for the $MgTiO_3$ are given in Equation (11).

$$\alpha_a = 8.4 \times 10^{-6} + 2.0 \times 10^{-9}\ (T - 25)\ [1/°C]$$
$$\alpha_c = 1.1 \times 10^{-5} + 1.7 \times 10^{-9}\ (T - 25)\ [1/°C] \quad (11)$$
$$\gamma = 2.8 \times 10^{-5} + 6.5 \times 10^{-9}\ (T - 25)\ [1/°C]$$

TECs obtained from previous sol-gel product at a temperature range between 700 and 1300 °C [4] (using Table 6) are given in Equation (12).

$$\alpha_a = 8.6 \times 10^{-6} + 1.8 \times 10^{-9}\ (T - 25)\ [1/°C]$$
$$\alpha_c = 1.3 \times 10^{-5} + 9.510^{-10}\ (T - 25)\ [1/°C] \quad (12)$$
$$\gamma = 2.1 \times 10^{-5} + 3.0{*}10^{-9}\ (T - 25)\ [1/°C]$$

Thermal expansion expression for the sol-gel derived products are calculated by combining the present and previous HT-XRD data [4] (Tables 5 and 6) with the whole temperature range (25–1300 °C), and the overall thermal expansion coefficients for $MgTiO_3$ are given in Equation (13).

$$\alpha_a = 8.5 \times 10^{-6} + 1.9 \times 10^{-9}\ (T - 25)\ [1/°C]$$
$$\alpha_c = 1.1 \times 10^{-5} + 1.7 \times 10^{-9}\ (T - 25)\ [1/°C] \quad (13)$$
$$\gamma = 2.8 \times 10^{-5} + 6.0 \times 10^{-9}\ (T - 25)\ [1/°C]$$

Thermal expansion expressions for solid-state reaction products calculated from HT-ND studies between 23 and 1212 °C of $MgTiO_3$ made by solid-state reaction [7,8] are given in Equation (14).

$$\alpha_a = 9.4 \times 10^{-6} + 1.8 \times 10^{-9}\ (T - 25)\ [1/°C]$$
$$\alpha_c = 1.2 \times 10^{-5} + 1.2 \times 10^{-9}\ (T - 25)\ [1/°C] \quad (14)$$
$$\gamma = 3.1 \times 10^{-5} + 4.0 \times 10^{-9}\ (T - 25)\ [1/°C]$$

3.5. Atomic Positions

Selected data of the refined atomic positions for geikielite as refined by the Rietveld method, are given in Table 7.

Table 7. Atomic positions for $MgTiO_3$ in sample GQ as function of temperature (hot-stage).

Sample		GQ (AR)	GQ25	GQ200	GQ400	GQ600	GQ800	GQ900	GQPOST
Position	**T [°C]**	**25**	**25**	**200**	**400**	**600**	**800**	**890**	**25**
c	z(Mg)	0.3556	0.3562	0.3563	0.3563	0.3571	0.3569	0.3574	0.3565
c	z(Ti)	0.1446	0.1444	0.1445	0.1445	0.1447	0.1450	0.1451	0.1453
f	x(O)	0.3173	0.3198	0.3203	0. 3199	0. 3193	0. 3142	0. 3169	0.3166
f	y(O)	0.0215	0.0214	0.0218	0. 0211	0. 0206	0. 0183	0.0190	0.0252
f	z(O)	0.2465	0.2433	0.2433	0. 2433	0. 2433	0. 2457	0. 2426	0.2465
	R_p	14.3	9.89	9.66	8.32	8.12	8.01	7.97	4.93
	R_{wp}	19.8	11.88	11.52	9.74	9.35	9.26	9.05	2.1
	χ^2	1.26	1.34	1.31	1.29	1.37	1.26	1.26	7.1

After cooling to room temperature, the sample returned into the initial as-received RT crystallographic data of qandilite and geikielite. The lattice parameters were in (Å): $MgTi_2O_4$ (qandilite) a = 8.4404 (6) and $MgTiO_3$ (geikielite): a = 5.054 (1); c = 13.905 (2). The atomic positions for the geikielite are given in Table 7 (column GQPOST).

4. Discussion

Figure 2 shows that the lattice parameters versus temperature of the reference material, Mg_2TiO_4, are in excellent agreement with the published HT-ND data [5]. Figure 4 shows a comparison between calculated lattice parameters versus temperature derived from Equations (9) and (10) showing complete overlapping of the lattice parameters of the reference material (Mg_2TiO_4) (Equation (9)) and published HT-ND study of Mg_2TiO_4 [5].

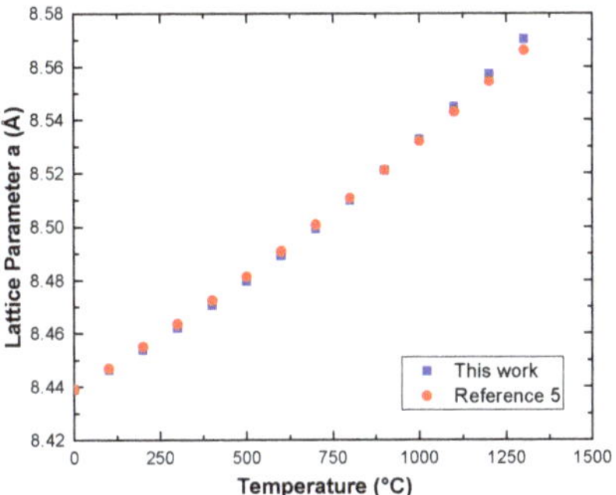

Figure 4. Calculated lattice parameters of Mg_2TiO_4 versus temperature, derived from Equations (9) and (10).

A comparison of Tables 5 and 6 shows that in $MgTiO_3$ there was no significant change in the atomic positions along the experimental temperature range (25–890 °C).

There was excellent agreement with present and previous HT-XRD data of a sol-gel $MgTiO_3$ product after 1 h firing [4] (Figure 3). The agreement with former HT-XRD on sol-gel products should be appreciated because, in contrast to usual TEC investigations, which are done on a single sample, it was done with different sample at each temperature. In order to eliminate the experimental scattering, the calculated lattice parameters versus temperature were plotted. Figure 5 shows that there was fair agreement between the lattice parameters versus temperature, between RT and 1300 °C, of all the sol-gel products of $MgTiO_3$ studied in HT-XRD data (present work and Reference [4]) in comparison with published results of minor $MgTiO_3$ phase in reference [7,8].

However, the lattice parameters of the sol-gel products were slightly lower. As presented in the literature [1,2], there is an asymmetric solubility gap in the geikielite with dissolving some amount of Ti at elevated temperatures. The lattice parameters of sample GQ ($MgTiO_3$ with a small amount of Mg_2TiO_4) fit those of stoichiometric MgTiO3 as reported in Reference [4]. This supports the phase diagram [1,2] that there is no Mg solubility. Since sample Kar2 [7,8] was a mixture of small amounts of $MgTiO_3$ and TiO_2 with $MgTi_2O_5$ as a major compound, it is reasonable to attribute the slight decrease of the lattice parameter of $MgTiO_3$ as stated in Reference [8] to some excess of Ti. In order to verify this hypothesis, we conducted an additional HT-XRD study of a second sol-gel product with a mixture of $MgTiO_3$ and $MgTi_2O_5$. The xerogel with $1 < Mg{:}Ti < 2$ was annealed 5 h at 1200 °C forming 78 wt% geikielite and 22 wt% karrooite. We designated this sample as "GK".

Figure 5 shows that at elevated temperatures, the lattice parameters of the $MgTiO_3$ as obtained from HT-XRD were slightly higher than those in sample GQ and in reference [4]. Moreover, they fit very well with those of sample Kar2 [7,8]. This confirms our assumption that the gap between the lattice parameters originated from excess of Ti in the Kar2 geikielite sample from references [7,8].

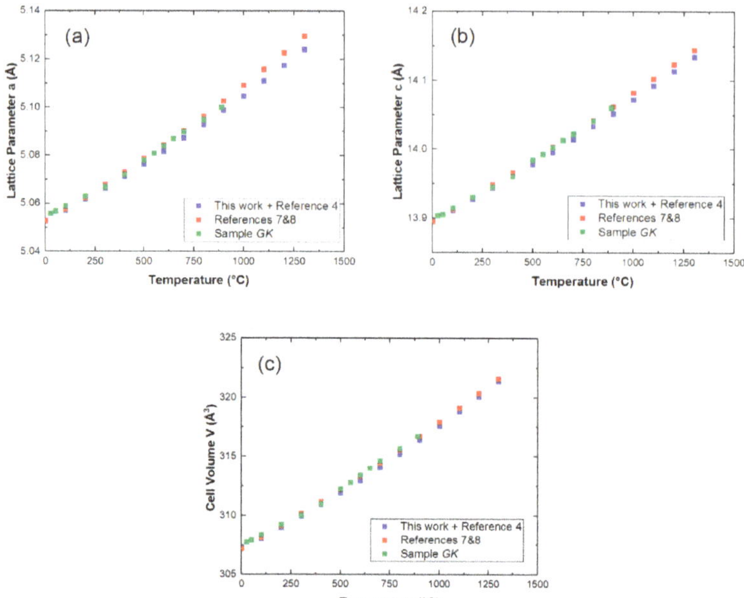

Figure 5. (**a**) Calculated lattice parameter a of $MgTiO_3$ versus temperature, derived from Equations (13) and (14) and experimental lattice parameter a of geikielite from sample GK. (**b**) Calculated lattice parameter c of $MgTiO_3$ versus temperature, derived from Equations (13) and (14) and experimental lattice parameter c of geikielite from sample GK. (**c**) Calculated lattice volume of $MgTiO_3$ versus temperature, derived from Equations (13) and (14) and experimental lattice volume of geikielite from sample GK.

The difference between the lattice volumes for sample *GQ*+ Reference [4] and sample *GK*+kar2 versus temperature is given in Figure 6, which shows that the gap between the lattice parameters increased with temperature until 1000 °C and then slightly decreased.

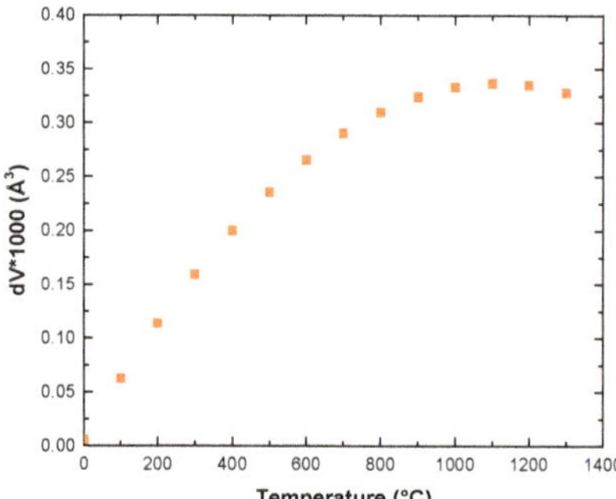

Figure 6. Difference between lattice volume of geikielite in GK sample in this work + references [7,8] and the GQ sample in this work+ reference [5] as function of temperature (using calculated data).

In this work, stoichiometric geikielite from eight samples made by sol-gel technique yielded new HT-XRD data. Therefore, as a result of the present work, it was found that there is no single set of TECs for the geikielite. It agrees with the phase diagram determined by Shindo [1] with an asymmetric solubility range in the geikielite. Both the sol-gel sample (GK) measured by HT-XRD and solid-state reaction (Kar2) measured by HT-ND were mixtures of geikielite and karrooite with maximum excess of Ti in the geikielite. Furthermore, both GK and Kar2 samples had similar lattice parameters, higher than the new data of the stoichiometric geikielite. Neither sample preparation nor diffraction method modified the TECs in geikielite. The fact that the sample GQ data integrated in seven stoichiometric samples [4] agrees with the phase diagram determined by Shindo [1] with absence of Mg solubility.

5. Conclusions

Accurate thermal expansion coefficients were measured for sol-gel products of stoichiometric $MgTiO_3$. The lattice parameters of $MgTiO_3$ made by sol-gel synthesis measured in HT-XRD between 25 and 890 °C are well integrated with the previously reported HT-XRD study of sol-gel $MgTiO_3$ product between 700 and 1300 °C. The lattice parameters of stoichiometric $MgTiO_3$ sol-gel products are slightly lower than nonstoichiometric $MgTiO_3$ with maximal excess of Ti. It is assumed the TECs of geikielite depend on deviations from stoichiometry. Neither sample preparation nor diffraction method modified the TECS in geikielite.

Author Contributions: T.T., student coordinator, literature survey, cross-checking all XRD results; G.K., R.T., XRD analysis of all XRD by Fullprof; B.A.R., collected the high-temperature XRD and performed the Rietveld analysis in Topas; J.Z., samples' production by sol-gel technique; H.D. and R.Z.S. synthesis of samples. All authors have read and agreed to the published version of the manuscript.

Funding: This research received no external funding.

Acknowledgments: The authors wish to acknowledge Elena Goncharov for the contribution to the synthesis.

Conflicts of Interest: The authors declare no conflict of interest.

References

1. Shindo, I. Determination of the phase diagram by the slow cooling float zone method: The $MgO-TiO_2$ system. *J. Cryst. Growth* **1980**, *50*, 839–851. [CrossRef]
2. Wechsler, B.A.; Navrotsky, A. Thermodynamics and structure chemistry of compounds in the system $MgO-TiO_2$. *J. Solid State Chem.* **1984**, *55*, 165–180. [CrossRef]
3. Wechsler, B.A.; Von Dreele, R.B. Structure refinements of Mg_2TiO_4, $MgTiO_3$, $MgTi_2O_5$ by time to flight neutron powder diffraction. *Acta Cryst.* **1989**, *B45*, 542–549. [CrossRef]
4. Zabicky, J.; Kimmel, G.; Goncharov, E.; Guirado, F. Magnesium titanate phases from xerogels by hot stage X-ray powder Diffractometry. *Z. Krist. Suppl.* **2009**, *30*, 347–352. [CrossRef]
5. O'Neill, H.S.C.; Redfern, S.A.T.; Kesson, S.; Short, S.M. An in situ neutron diffraction study of cation disordering in synthetic qandilite Mg_2TiO_4 at high temperatures. *Am. Mineralog.* **2003**, *88*, 860–865. [CrossRef]
6. Tuval, T.; Zabicky, J.; Gispert-Guiardo, G.; Kimmel, G.; Gomcharoc, E.; Shneck, R.Z. In situ HTXRD formation of magnesium titanates. *J. Am. Ceram. Soc.* **2018**, *101*, 4367–4374. [CrossRef]
7. Lennie, A.R.; Knight, K.S.; Michael, C.; Henderson, B. Cation ordering in $MgTi_2O_5$ (karrooite): Probing temperature dependent effects with neutrons. *Am. Mineral.* **2007**, *92*, 165–1180. [CrossRef]
8. Henderson, C.M.B.; Knight, K.S.; Lennie, A.R. Temperature dependence of rutile (TiO_2) and geikielite ($MgTiO_3$). *Open Mineral. J.* **2009**, *3*, 1–11. [CrossRef]
9. Rodriguez-Carvajal, J. Recent advances in magnetic structure determination by neutron powder diffraction. *Phys. B Condens. Matter* **1993**, *192*, 55–69. [CrossRef]
10. Kraus, K.; Nolze, G. Powder cell—A program for the representation and manipulation of crystal structures and calculation of the resulting X-ray powder patterns. *J. Appl. Cryst.* **1996**, *29*, 301–303. [CrossRef]

Article

Effect of Various Nanoparticles (GaF_3, ZnF_2, $Zn(BF_4)_2$ and Ga_2O_3) Additions on the Activity of CsF-RbF-AlF_3 Flux and Mechanical Behavior of Al/Steel Brazed Joints

Zhen Yao, Songbai Xue * and Junxiong Zhang

College of Materials Science and Technology, Nanjing University of Aeronautics and Astronautics, Nanjing 210016, China; yaozhen123@nuaa.edu.cn (Z.Y.); zhangjunxiong126@163.com (J.Z.)
* Correspondence: xuesb@nuaa.edu.cn; Tel.: +86-025-8489-6070

Received: 15 July 2020; Accepted: 5 August 2020; Published: 7 August 2020

Abstract: In this study, brazing AA6061 to Q235 steel using flame brazing was performed with 70.9 wt.% CsF-0.5 wt.% RbF-28.6 wt.% AlF_3 fluxes doped with GaF_3, ZnF_2, $Zn(BF_4)_2$ and Ga_2O_3 nanoparticles, matched with a Zn-15Al filler metal, and the spreadability of the filler metal and the mechanical properties of brazed joints were investigated at the same time. The results showed suitable amounts of GaF_3, ZnF_2, $Zn(BF_4)_2$ and Ga_2O_3 doped into the base flux could strengthen the filler metal in wetting and spreading on the surface of aluminum alloy and steel to different degrees. The suitable ranges of GaF_3, ZnF_2, $Zn(BF_4)_2$ and Ga_2O_3, respectively, were 0.0075–0.01 wt.%, 0.0075–0.01 wt.%, 0.0075–0.01 wt.% and 0.009–0.01 wt.%, and the maximum spreading area was obtained via doping with GaF_3. The shear strength of brazed joints reached the peak at 126 MPa when 0.075 wt.% GaF_3 was added. Comparative tests proved that the activity of the CsF-RbF-AlF_3 flux doped with GaF_3 was the best. The reason was that the CsF-RbF-AlF_3-GaF_3 flux was competent in removing oxides of the base metal and decreasing the interfacial tension, in virtue of the activity of Ga^{3+} as well as F^-.

Keywords: GaF_3; ZnF_2; $Zn(BF_4)_2$; Ga_2O_3; flux; Zn-Al filler metal; spreadability; mechanical properties

1. Introduction

The increase in joining dissimilar metals has been prompted by the increasing demand for light weights and fuel efficiency in modern industries in recent years [1]. The emerging trend towards lightweight, high performance and emissions reduction is leading to the increasing use of multi-material hybrid structures in electric vehicles [2]. The hybrid structure of aluminum and steel are superior to conventional materials because of the high mechanical properties of steel and the excellent corrosion resistance of aluminum [3]. Naturally, the combination of aluminum alloy and steel shows huge research value and future potential, and has become the subject of many investigators [4].

However, it remains an unconquerable problem to get stable brazed joints between aluminum/metals dissimilar to steel because of the formation of brittle and hard intermetallic compounds, like Fe_xAl_y [5]. The complexity and randomness of the brazing process makes the design and development of brazing materials more complicated and time-consuming than ordinary materials [6]. Therefore, it is absolutely imperative to control the formation of Fe-Al intermetallic compounds. EI-Sayed and Naka [7] found that the maximum bond shear strength of 127 MPa was obtained for an Al-steel joint brazed at 663 K with a 3 s ultrasound application time using a Zn-14Al alloy. It was reported that the shear strength of brazed joints reached the peak at 131 MPa when the Zn–15Al filler metal was added in the joining of lap joints of 6061 aluminum alloy to the 304 stainless steel via a flame brazing process, with the Zn-xAl filler metals matching the CsF-0.5 wt.%

$RbF-AlF_3$ flux [8]. The interfacial layer in the weld made with the Zn-15Al filler metal was comprised of $(FeAl_3)Zn_x$ and $(Fe_2Al_5)Zn_x$ [9].

Nanomaterials have been widely applied due to their unique properties. Joining technology at the nanometer scale has gradually developed with the popularity of nanomaterials, which have broad application prospects in the fields of electronics, aero-space, biology and health care [10]. Previous studies have indicated that [11] trace amounts of Ga_2O_3 addition in the CsF-0.5 wt.% $RbF-AlF_3$ flux could obviously strengthen the Zn-2Al filler metal in its wetting and spreading on the surface of 5052 aluminum alloy and Q235 low-carbon steel. One previous study discussed how the addition of GaF_3 and Ga_2O_3 nanoparticles influenced the wettability and spreadability of the $CsF-AlF_3$ flux under the same conditions [12].

The influence that adding heavy metal fluoride, ZnF_2, SnF_2, CdF_2, PbF_2 and KBF_4 into the $KF-AlF_3$ flux has on the spreadability of brazing aluminium was investigated, and adding ZnF_2 could greatly improve the brazing area [13]. Therefore, the fourth component, ZnF_2, has been considered for doping into CsF-0.5 wt.% $RbF-AlF_3$ flux in order to reduce the price in this paper. $ZnCl_2$ and $SnCl_2$ were added into the $CsF-AlF_3$ flux for connecting aluminium alloys [14]. The joints were bonded soundly when the mass fractions of $ZnCl_2$ and $SnCl_2$ are about 4%.

The effect of KBF_4 addition on the microstructure of the Mg-6Zn-1Si alloy has been investigated [15], and the morphology of the Mg_2Si phase changed with the addition of 1.5 wt.% KBF_4. To compare the activity of BF_4^- and its unified positive ions in reducing the variables, $Zn(BF_4)_2$ has been chosen.

In this paper, brazing AA6061 to Q235 steel using flame brazing has been performed with improved CsF-0.5 wt.% $RbF-AlF_3$ fluxes doped with a GaF_3, ZnF_2, $Zn(BF_4)_2$ and Ga_2O_3 nanoparticles-matched Zn-15Al filler metal, and the spreadability of the filler metal and the mechanical properties of the brazed joints were investigated at the same time. XDR analysis was carried out and the reaction mechanism was analyzed. The results could be useful for brazing AA6061 to Q235 steel while choosing a suitable flux.

2. Materials and Methods

The 6061 aluminum alloy and Q235 steel were used in this work as base metals. The compositions of the base metals were listed in Tables 1 and 2. Zn–15Al alloys were chosen as the filler metal. A CsF-0.5 wt.% $RbF-AlF_3$ flux was prepared by using commercial $CsF-AlF_3$ flux and an RbF of AR purity (Zhejiang Xinrui Welding Materials Co., Ltd., Shengzhou, China). $CsF-RbF-AlF_3-GaF_3$ fluxes with different compositions of AR purity GaF_3 were confected and the range of GaF_3 was 0.0001–0.125 wt.%. Same amounts of ZnF_2, and $Zn(BF_4)_2$ of AR purity were doped in the same way to obtain $CsF-RbF-AlF_3-ZnF_2$ fluxes and $CsF-RbF-AlF_3-Zn(BF_4)_2$ fluxes. Nano Ga_2O_3 powder in the same range was doped into the $CsF-RbF-AlF_3$ flux to get the corresponding $CsF-RbF-AlF_3-Ga_2O_3$ fluxes. After mixing into a liquid, the fluxes were then dried to powder with the oven. The chemical compositions of these four nanoparticles doped into the fluxes has been listed in Table 3.

Table 1. Chemical composition of 6061 aluminum alloy (wt.%).

Alloy	Mg	Si	Cu	Cr	Mn	Zn	Al
6061	1.10	0.61	0.25	0.12	0.01	0.01	Bal.

Table 2. Chemical composition of Q235 steel (wt.%).

Alloy	C	Mn	Si	S	P	Fe
Q235	0.18	0.48	0.30	0.04	0.04	Bal.

Table 3. Chemical composition of nanoparticles doped into the fluxes (wt.%).

Number	1	2	3	4	5	6	7	8	9	10	11
Composition	0	0.0001	0.001	0.00325	0.0045	0.005	0.006	0.0075	0.009	0.01	0.0125

Zn–15Al alloys were extruded into a 2 mm diameter wire as an advanced preparation. SiC paper was used for mechanically polishing the specimens and filler metals. Before brazing, these materials were all degreased with acetone and cleaned by ethanol.

In order to prepare for the spreading test, the base metals were processed into plates of 40 mm × 40 mm × 3 mm. The spreading test was performed according to China's National Standard GB 11364-2008. The weights of the solder and flux used in the test should be 100 mg and 15 mg, respectively. The filler metals were placed in the center of the base metals covered with the fluxes prepared previously, as in Figure 1, and then put into the electrical resistance furnace (Zhejiang Xinrui Welding Materials Co., Ltd., Shengzhou, China). The heating temperature was uniformly set at 530 °C. The holding time of each test was set as 60 s. After spreading, the test boards were cleaned using ultrasonic wave waits and the spreading areas were calculated by the software Image-ro Plus (Image-Pro Plus Version 6.0). To be specific, we photographed the spreading boards with the graduated ruler and imported the image into Image-Pro Plus. By determining the actual scale in the picture and using chromatic aberration to circle the spreading outline, the spreading areas could be calculated. For the credibility of the results, the above tests of each group were repeated 5 times and the spreading areas of each group were averaged. At last, the residues of the fluxes were collected and the components in the residues were analyzed with a Brucker D8 XRD analyzer (Ningbo Institute of Materials Technology & Engineering, Ningbo, China).

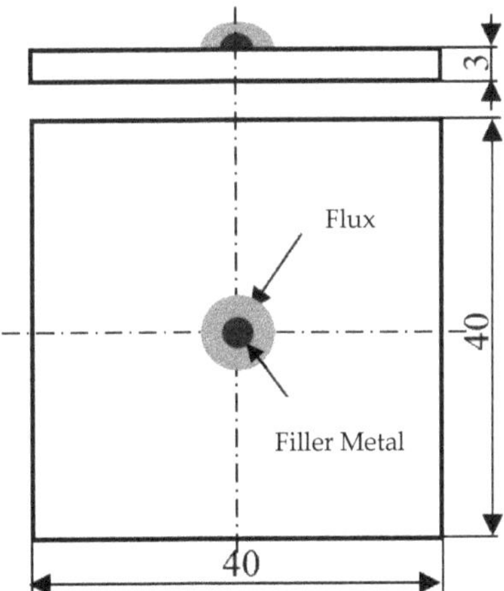

Figure 1. Schematic diagram of the spreading test (mm).

For the shear performance tests, the supplied base metals for the brazed joint were processed into plates with the size of 60 mm × 25 mm × 3 mm. Figure 2 shows the schematic illustration of the brazed joint. The shear performance tests of the AA6061/Q235 brazed joints were carried out in strict accordance with China's National Standard GB 11363-2008. The equipment was an SAMS-CMT5105 universal tensile testing machine (Nanjing University of Aeronautics and Astronautics, Nanjing, China), and the loading rate in the tensile process was 3 mm/min. The test results for each group of samples were all averaged from 5 samples. The brazing temperature was detected and controlled at around 530 °C.

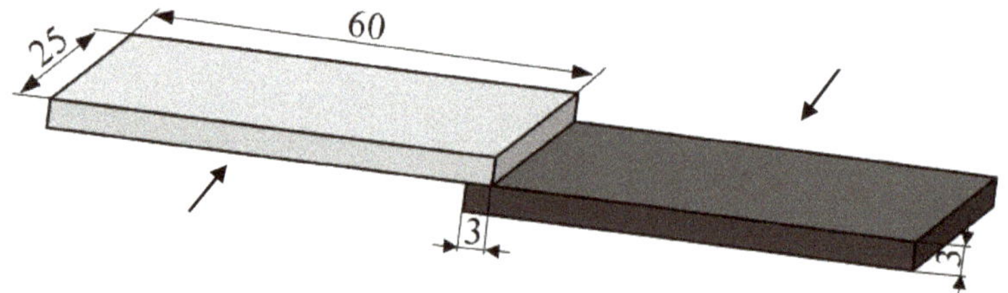

Figure 2. Schematic illustration of the brazed joint (mm).

3. Results and Discussion

3.1. The Spreadability and Wettability of Zn-15Al Filler Metal

The influence of GaF_3, ZnF_2, $Zn(BF_4)_2$ and Ga_2O_3 particles, doped into CsF-0.5 wt.% RbF-AlF_3 fluxes, on the spreadability of the Zn-15Al filler metal was studied via spreading tests performed on the surface of the 6061 aluminum alloy and the Q235 steel. The results were as follows.

The relationships between the concentrations of GaF_3, ZnF_2, $Zn(BF_4)_2$ and Ga_2O_3 and the spreading areas on the 6061 aluminum alloy are shown in Figure 3a. While the content of GaF_3 was 0.01 wt.%, the spreading area was maximized to 329 mm^2, which was 65% larger than the area without any addition. With 0.0075 wt.%-doped ZnF_2, the spreading area was maximized at 312 mm^2, and showed a 56% increase compared with the results derived without adding ZnF_2. While the content of $Zn(BF_4)_2$ was 0.0075 wt.%, the spreading area was maximized at 304 mm^2, and was 52% larger than the area without any addition. With 0.009 wt.%-doped Ga_2O_3, the spreading area was maximized at 321 mm^2 and showed a 61% increase compared with the results derived without adding Ga_2O_3.

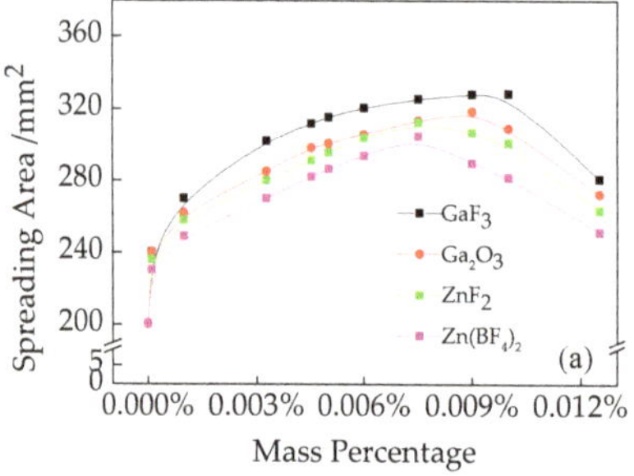

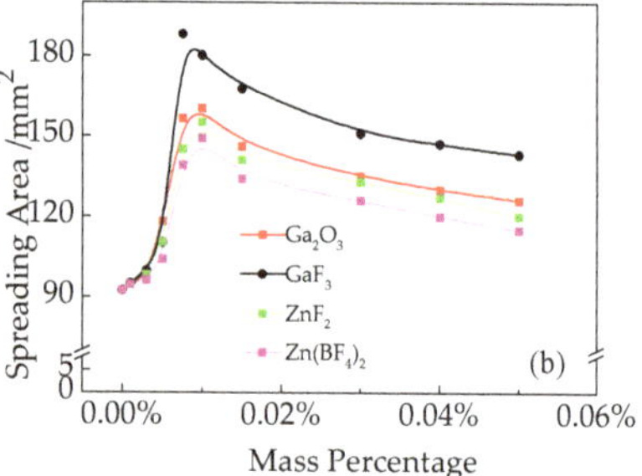

Figure 3. Spreading areas of Zn-15Al filler metal (**a**) 6061 aluminum alloy; (**b**) Q235 steel.

Figure 3b showed that the spreadability of the Zn-15Al filler metal over Q235 steel was clearly improved with GaF_3 doped into the CsF-RbF-AlF_3 flux, compared to others. While the content of GaF_3 was 0.0075 wt.%, the spreading area was maximized at 189 mm^2. With 0.01 wt.%-doped ZnF_2, the spreading area was maximized at 155 mm^2. While the content of $Zn(BF_4)_2$ was 0.01 wt.%, the spreading area was maximized at 149 mm^2. With 0.01 wt.%-doped Ga_2O_3, the spreading area was maximized at 161 mm^2. With the addition of GaF_3, the spreading areas over Q235 steel underwent a 105% increase, compared with the results derived without adding GaF_3 (92 mm^2), and this substance showed clear advantages over ZnF_2, $Zn(BF_4)_2$ and Ga_2O_3.

The pictures of the best spreadings of Zn-15Al alloys on AA6061 and Q235, with CsF-RbF-AlF_3 flux doped with different additions, are shown in Figure 4. Taking the test results from two kinds of base metal together, it could be concluded that the suitable ranges of GaF_3, ZnF_2, $Zn(BF_4)_2$ and Ga_2O_3 in the CsF-RbF-AlF_3 flux respectively were 0.0075–0.01 wt.%, 0.0075–0.01 wt.%, 0.0075–0.01 wt.% and 0.009–0.01 wt.%, and the maximum spreading area of all the tests was obtained via doping with GaF_3.

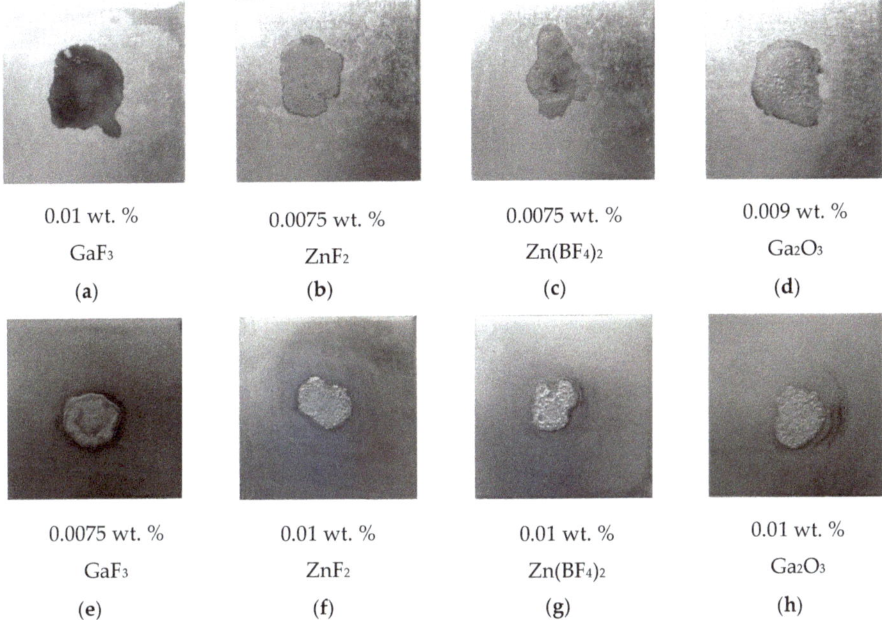

Figure 4. Spreading boards of Zn-15Al filler metal on base metals with CsF-RbF-AlF_3 flux doped with different additions: (**a–d**) 6061 aluminum; (**e–h**) Q235 low-carbon steel.

3.2. The Mechanical Properties of Brazed Joints

The variation of brazed joint shear strength was shown in Figure 5. It could be seen that the addition of GaF_3 has a significant impact on the brazed joint shear strength of AA6061/Q235. While the content of GaF_3 was 0.075 wt.%, the shear strength reached a maximum at 126 MPa, which was 110% higher than that achieved without GaF_3 addition (64.5 MPa). While the content of ZnF_2 was 0.01 wt.%, the maximum shear strength was 111 MPa. While the content of $Zn(BF_4)_2$ was 0.01 wt.%, the shear strength reached the top at 105 MPa. While the content of Ga_2O_3 was 0.01 wt.%, the maximum shear strength was 116 MPa.

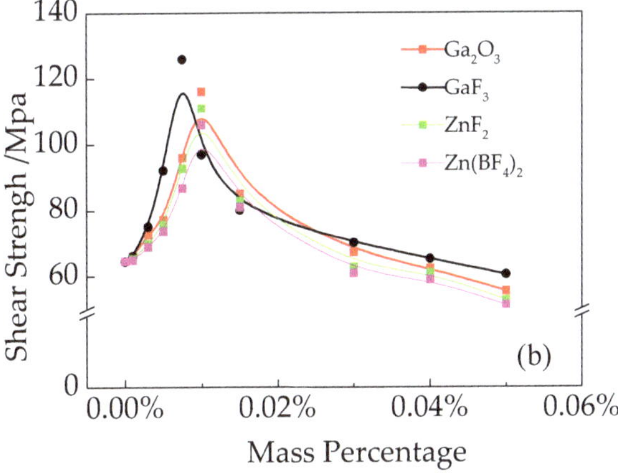

Figure 5. Influence of contents of GaF_3, ZnF_2, $Zn(BF_4)_2$ and Ga_2O_3 on shear strength of brazed joints.

It could be seen that, only considering the shear strength of the AA6061/Q235 brazed joints, the addition range should be controlled as 0.0075–0.01 wt.%, and this appropriate addition range coincides with the results of the spreading test. Among the four doped ingredients selected, GaF_3 showed the best performance as regards the mechanical properties of brazed joints. The second-best performance was achieved with Ga_2O_3. At the same time, $Zn(BF_4)_2$ was inferior to others.

The typical fracture modes of the 6061/Q235 brazed joints were shown in Figure 6. The fracture of the joint occurred mainly at the interface layer of the brazing joint and the Q235 steel. It demonstrated that the interface layer between the filler metal and the Q235 steel was the weakest area of the whole joint, and that when there were layers of brittle compounds at the interface, the brazed joint would crack at the layers of brittle compounds first.

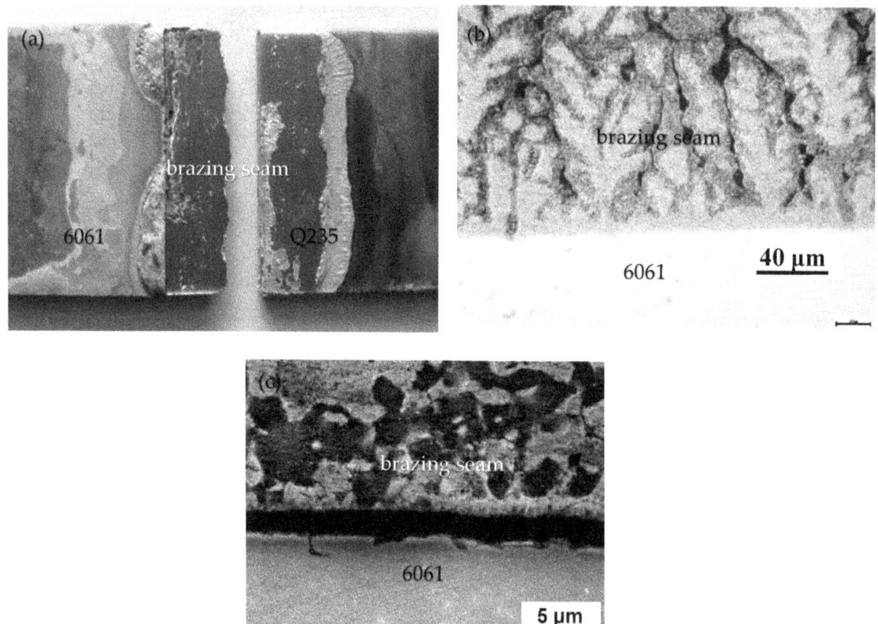

Figure 6. Typical fracture modes of 6061/Q235 brazed joints: (**a**) Macromorphology of fracture interface; (**b**) Organization of interface before shear tests; (**c**) Microtopography of fracture interface after shear tests.

3.3. Interfacial Effect of Different Particles

3.3.1. Effect of $Zn(BF_4)_2$

The XRD analysis results for the residues of the fluxes on Q235 low-carbon steel are shown in Figure 7. In inndividually evaluating the spreadability of the Zn-15Al filler metal, and the mechanical properties of the brazed joints with CsF-RbF-AlF_3-$Zn(BF_4)_2$ over the 6061 aluminum and Q235 steel, it was clear that the addition of $Zn(BF_4)_2$ significantly improved the activity of the CsF-RbF-AlF_3 flux, with a 52% increase in spreading area and a 92% increase in shear strength.

$$Zn(BF_4)_2 \rightarrow ZnF_2 + 2BF_3 \quad (1)$$

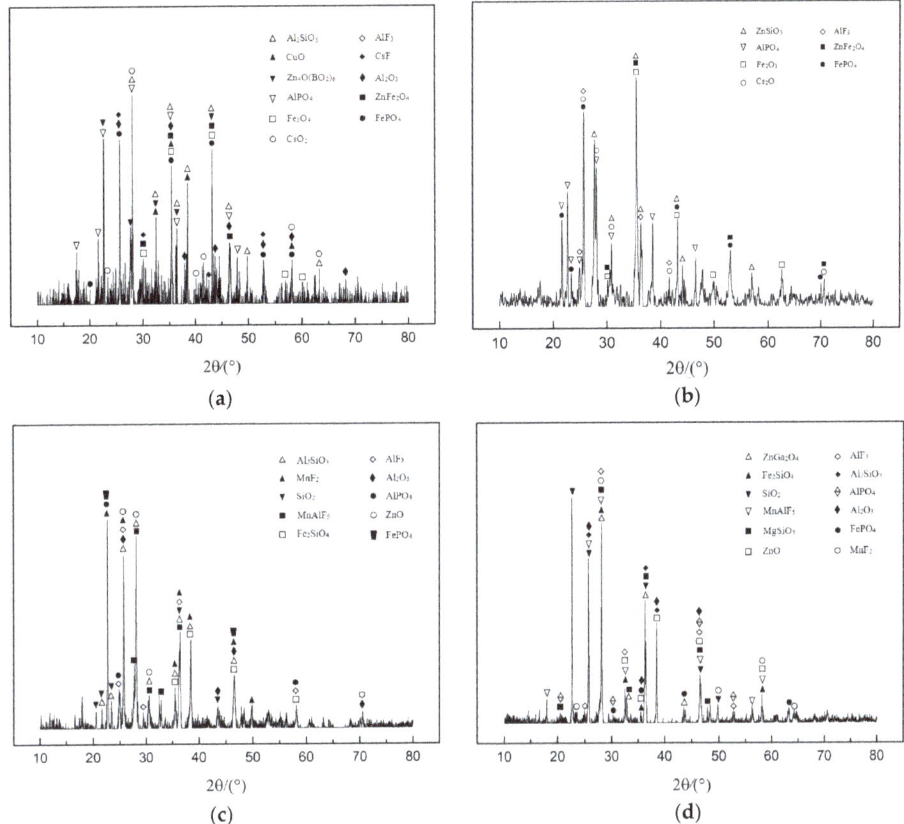

Figure 7. X-ray diffraction analysis results of the residues on Q235 low-carbon steel: (**a**) CsF-RbF-AlF_3-$Zn(BF_4)_2$; (**b**) CsF-RbF-AlF_3-ZnF_2 flux; (**c**) CsF-RbF-AlF_3- Ga_2O_3 flux; (**d**) CsF-RbF-AlF_3-GaF_3 flux.

$Zn(BF_4)_2$ could produce BF_3 as shown Equation (1). BF_3 could react with oxides such as FeO, Fe_2O, NiO, Cr_2O_3 and ZnO, thus playing the role of removing the oxide film and promoting the wetting, spreading and flowing of the molten filler metal in the brazed part [16]. However, BF_3 could not react with oxides such as Al_2O_3 and MgO on the surface of t he aluminum alloy, thus limiting the activity of the CsF-RbF-AlF_3-$Zn(BF_4)_2$ flux.

3.3.2. Effect of ZnF_2

It was found in the study [13] that adding ZnF_2 to the flux could produce a mass transfer effect, and improve the activity of the flux. This was because the ZnF_2 was reduced by the base metal, and molten Zn has great solubility in aluminum. It was also observed that adding ZnF_2 could reduce the initial temperature of the flux, which was also related to the reduction of the precipitation of molten Zn.

$$ZnF_2 \xrightarrow{\Delta} Zn \downarrow + 2F^- \quad (2)$$

$$F^- + Al_2O_3 \rightarrow AlF_3 + O^{2-} \quad (3)$$

$$F^- + Fe_2O_3 \rightarrow FeF_3 + O^{2-} \quad (4)$$

When heated, ZnF_2 would react as shown Equation (2). The F^- generated by the reaction would react with the oxide on the surface of the base metals, as shown in Equations (3) and (4), respectively, thus enhancing the membrane removal effect of CsF-RbF-AlF_3 flux. The Zn atoms precipitated from the reaction would react with the aluminum atoms on the surface of the aluminum alloy and the Fe atoms on the surface of steel, and the wetting and spreading would be promoted on the surface of the base metals [17].

The presence of $ZnSiO_3$, $ZnFe_2O_4$, $AlPO_4$ and $FePO_4$ in the residue on the one hand indicated that the ZnF_2 in the flux did participate in the reaction, and the Zn atoms from the reduced precipitation participated in the reaction and played a role in improving the activity. On the other hand, it indicated that the enrichment phenomenon of P occurred, which resulted in the formation of compounds such as $AlPO_4$ and $FePO_4$, which also played an important role in the removal of the oxide film from the steel surface.

3.3.3. Effect of Ga_2O_3

The "skin effect" of high-frequency currents is a well-known natural phenomenon in physics, especially in electromagnetism. However, the "skin effect" of some oxides or halides in brazing has rarely been reported. It was found that [18] adding a very small amount of Ga_2O_3 to the CsF-RbF-AlF_3 flux could increase the brazed area by about 50~90%. Observing the surface of the spreading filler metal via its spectrum showed that much Ga_2O_3 was enriched at the spreading surface and at the edge of the spreading area. Further, the relative amount of Ga_2O_3 was higher here than in the middle part, from which it could be concluded that the chemical reaction mechanism resulted from the skin effect of the Ga_2O_3, which enabled the Ga_2O_3 to flow fast and then promote the CsF-RbF-AlF_3 flux flow, so as to improve the spreadablity of the filler metal.

$$MgO + Ga_2O_3 \rightarrow MgGa_2O_4 \tag{5}$$

Extremely small amounts of Ga_2O_3 were enriched to participate in the reaction, and played a nonnegligible role. The appearance of $MgGa_2O_4$ indicated that the reaction between the flux and the MgO, which was more stable than Al_2O_3, was also liable to react as in Equation (5). However, RbF did not appear as a compound phase in the residue, which indicated that the compound of Rb was diffused-distributed, and acted as catalysis and welding aid. CsF always appeared in the form of $Cs_{11}O_3$, which indicated that CsF did undergo chemical reactions with Al_2O_3 and MgO, and replaced the O atoms in Al_2O_3 and MgO. This result explains why the spreading area increased with the increased addition of Ga_2O_3, shown in Figure 3.

3.3.4. Effect of GaF_3

From the above it was demonstrated that the CsF-RbF-AlF_3 flux doped with GaF_3 obtained the maximum spreading area, both on 6061 aluminum alloy and Q235 steel, and reached the highest shear strength of 126MPa. The CsF-RbF-AlF_3 flux doped with GaF_3 showed remarkable superiority to that doped with ZnF_2, $Zn(BF_4)_2$ and Ga_2O_3.

The mechanism of GaF_3 could be summarized as a "synergistic effect" on the oxide removal of steel. When GaF_3 was doped, the flux could not only remove the oxide films on the surface of the aluminum alloy, but could also remove those on the surface of the steel at the same time, thus making it superior to brazed aluminum-steel heterogeneous materials.

$$GaF_3 \rightarrow Ga^{3+} + 3F^- \tag{6}$$

$$4ZnO + 2GaF_3 \rightarrow ZnGa_2O_4 + 3ZnF_2 \tag{7}$$

The ionic compound GaF_3 consists of cations Ga^{3+} and anions F^-, and makes it such that Ga^{3+} and F^- ions can be more easily dissociated from GaF_3 than from Ga_2O_3, as in Equation (6), while been heated. This flux was competent in removing the oxides of the base metal and decreasing the interfacial tension, in virtue of the activity of Ga^{3+}, which has "skin effect", as well as F^-, which reacts as in Equations (3) and (4). In addition, as Equation (7) showed, the production of ZnF_2 enhanced the activity of the flux.

4. Conclusions

It was shown that the CsF-RbF-AlF_3 flux doped with GaF_3 obtained the maximum spreading area, both on 6061 aluminum alloy and Q235 low-carbon steel, and reached the highest shear strength of 126MPa. The above analysis led to the following conclusions:

(1) The spreading tests indicated that the spreadability of the Zn-15Al filler metal both on Q235 steel and AA6061 alloy was promoted with a CsF-RbF-AlF_3 flux doped with GaF_3, ZnF_2, $Zn(BF_4)_2$ and Ga_2O_3, and the performance of GaF_3 was the best. The suitable ranges of GaF_3, ZnF_2, $Zn(BF_4)_2$ and Ga_2O_3 respectively were 0.0075–0.01 wt.%, 0.0075–0.01 wt.%, 0.0075–0.01 wt.% and 0.009–0.01 wt.%.

(2) The shear strength of the brazed joints reached its peak at 126 MPa, when 0.075 wt.% GaF_3 was added. The second-best performance was Ga_2O_3. At the same time, $Zn(BF_4)_2$ was inferior to others.

(3) The fluxes doped with GaF_3 were competent in removing oxides of the base metal and decreasing the interfacial tension, in virtue of the activity of Ga^{3+}, which has a "skin effect", as well as F^-. In consequence, the activity of the CsF-RbF-AlF_3 doped with GaF_3 was better than that of others.

Author Contributions: Conceptualization, S.X.; methodology, J.Z.; software, Z.Y.; validation, Z.Y.; formal analysis, Z.Y.; investigation, Z.Y.; resources, S.X.; data curation, Z.Y.; writing—original draft preparation, Z.Y.; writing—review and editing, Z.Y.; visualization, Z.Y.; supervision, S.X.; project administration, S.X.; funding acquisition, S.X. All authors have read and agreed to the published version of the manuscript.

Funding: This work was funded by the National Natural Science Foundation of China, Grant No. 51375233 and the Priority Academic Program Development of Jiangsu Higher Education Institutions (PAPD).

Acknowledgments: We gratefully acknowledge the support of Jiangsu Higher Education Institutions for providing a scholarship for Zhen Yao. Songbai Xue acknowledges financial support from the National Natural Science Foundation of China, Grant No. 51375233.

Conflicts of Interest: The authors declare no conflict of interest.

References

1. Li, P.; Lei, Z.L.; Zhang, X.R.; Liu, J.G.; Chen, Y.B. Effects of laser power on the interfacial intermetallic compounds and mechanical properties of dual-spot laser welded–brazed Ti/Al butt joint. *Opt. Laser Technol.* **2020**, *124*, 1–11. [CrossRef]
2. Li, H.; Liu, X.S.; Zhang, Y.S.; Senkara, J.; Li, G.Y.; Ma, M.T. Current research and challenges in innovative technology of joining dissimilar materials for electric vehicles. *Mater. Rep.* **2019**, *33*, 3853–3861.
3. Xue, S.B.; Wang, B.; Zhang, L.; Long, W.M. Development of green welding technology in China during the past decade. *Mater. Rep.* **2019**, *33*, 2813–2830.
4. He, P.; Wang, H.Q.; Feng, J.C.; Qian, Y.Y. Optimization of induction coil of high frequency brazing broad plane aluminium to stainless steel. *Weld. Join.* **2005**, *5*, 20–23.
5. Matsuda, T.; Sano, T.; Munekane, M.; Ohata, M.; Hirose, S. Microscale tensile test of galvannealed steel/aluminum dissimilar joint for estimation of intrinsic interfacial strength. *Scr. Mater.* **2020**, *186*, 196–201. [CrossRef]
6. He, P.; Lin, P.P. The systematic project involving brazes development and intelligent brazing technology innovation: A materials genome perspective. *Mater. Rep.* **2019**, *01*, 156–161.
7. El-Sayed, M.H.; Naka, M. Structure and properties of carbon steel–aluminium dissimilar joints. *Sci. Technol. Weld. Join.* **2005**, *10*, 27–31. [CrossRef]
8. Yang, J.L.; Xue, S.B.; Zhang, J.X. Development of novel CsF–RbF–AlF_3 flux for brazing aluminum to stainless steel with Zn–Al filler metal. *Mater. Des.* **2014**, *64*, 110–115. [CrossRef]

9. Dong, H.G.; Yang, L.Q.; Zhai, N.; Dong, C. Analysis on microstructure and mechanical properties of aluminum alloy/stainless steel joint made with flux-cored filler metal. *Trans. China Weld. Inst.* **2011**, *32*, 1–4.
10. He, P.; Jiao, Z.; Wang, J.; Lin, T.S. Research and application of joining technology at nanometer scale. *Trans. China Weld. Inst.* **2013**, *34*, 109–112.
11. Yao, Z.; Xue, S.B.; Yang, J.L.; Zhang, J.X. Inducing the effect of a Ga_2O_3 nano-particle on the CsF-RbF-AlF_3 flux for brazing aluminum to carbon steels. *Crystals* **2020**, *10*, 183. [CrossRef]
12. Yao, Z.; Xue, S.B.; Zhang, J.X. Comparative study on the activity of GaF_3 and Ga_2O_3 nanoparticle-doped CsF-AlF_3 flux for brazing 6061 Al/Q235 steel joints. *Crystals* **2020**, *10*, 498. [CrossRef]
13. Zhang, L.J. Investigation to fluoride brazing flux used for aluminium brazing. *Weld. Join.* **1996**, *6*, 7–8.
14. Cheng, F.J.; Qi, S.M.; Yang, Z.W.; Yao, J.F.; Zhao, H. Self-brazing mechanism of aluminum alloy at medium temperature. *J. Mater. Eng.* **2018**, *46*, 31–36.
15. Zhang, B.F.; Chen, Q.; Chen, X.W.; Zhang, H.L. Effect of KBF_4 addition on the microstructure of Mg-6Zn-1Si alloy. *Adv. Mater. Res.* **2012**, *535*, 996–999. [CrossRef]
16. Xiao, D.F.; Chen, D.H.; Qian, Y.Y.; Mao, J.Y.; Fan, F.H.; Yan, J. Study on corrosion resistance of aluminum soldering joint. *Weld. Join.* **1985**, *09*, 3–6.
17. Zhang, Q.Y.; Liu, S.Q.; Lan, T.; Fu, Y.C. Interficial activity behavior of flux during aluminum brazing. *Acta Metall. Sin.* **1989**, *4*, 137–141.
18. Zhang, J.X.; Xue, S.B.; Xue, P.; Yang, J.L.; Lv, Z.P. Effect of Ga_2O_3 on the wettability and spreadability of CsF-AlF_3 flux/Zn-Al filler metal on the aluminum and steel. *Rare Met. Mater. Eng.* **2017**, *7*, 156–160.

Article

Comparative Study on the Activity of GaF_3 and Ga_2O_3 Nanoparticle-Doped CsF-AlF_3 Flux for Brazing 6061 Al/Q235 Steel Joints

Zhen Yao, Songbai Xue * and Junxiong Zhang

College of Materials Science and Technology, Nanjing University of Aeronautics and Astronautics, Nanjing 210016, China; yaozhen123@nuaa.edu.cn (Z.Y.); zhangjunxiong126@163.com (J.Z.)

* Correspondence: xuesb@nuaa.edu.cn; Tel.: +86-025-8489-6070

Received: 6 May 2020; Accepted: 8 June 2020; Published: 9 June 2020

Abstract: The effect of trace amounts of GaF_3 and Ga_2O_3 nanoparticles on the wettability and spreadability of CsF-AlF_3 flux matched Zn-15Al filler metal were comparatively studied on 6061 aluminum alloy and Q235 low-carbon steel. The experimental results indicate that appropriate amounts of GaF_3 and Ga_2O_3 added into the flux could significantly promote the Zn-15Al filler metal to wet and spread on the surface of 6061 aluminum alloy and Q235 low-carbon steel. The optimum ranges for GaF_3 and Ga_2O_3 were 0.0075–0.01wt.% and 0.009–0.01 wt.%, respectively. Comparative analysis showed that the activity of CsF-AlF_3 flux bearing GaF_3 was higher than that bearing Ga_2O_3. The reason for this is that the former flux has a stronger ability to remove oxides of the base metal and reduce the interfacial tension of the molten filler metal and the base metal.

Keywords: spreadability; wettability; GaF_3; Ga_2O_3; CsF-AlF_3 flux

1. Introduction

Considering their excellent physical and chemical properties as well as good mechanical processability and corrosion resistance, aluminum alloys has become the most widely used non-ferrous metal structural materials in modern industries, including appliances, construction, light industry, storage tanks, aviation, aerospace, automobile and weapons. Among them, 6061 aluminum alloy has the advantages of excellent formability, welding performance, machinability and corrosion resistance [1,2]. Despite their outstanding properties, the mechanical strength of aluminum alloys is occasionally insufficient to meet the requirements of rigid structures such as car body keel, while the welding structure of aluminum and steel can effectively reduce the weight and additionally exhibit an enhanced strength. Therefore, the combination of aluminum alloy and steel shows great potential in the manufacturing industry.

Although the combination of aluminum alloy and steel plays an important role in the modern manufacturing industry, it remains a hot topic and difficult problem to obtain stable and reliable welded joints in the welding field, particularly for complicated situation in aluminum/steel dissimilar metals [3,4]. Domestic and foreign scholars were committed to the problem of aluminum steel welding in recent years and have proposed various welding approaches, such as cold metal transition brazing [5], aluminum alloy/galvanized steel MIG welding [6] and so on. The difficulties in the welding of aluminum/steel dissimilar metals is mainly related to the following two aspects: the different physical properties of aluminum and steel, which lead to considerable welding stress in the joints, and the brittle intermetallic compounds of the Fe-Al phase generated at the interface between aluminum and steel which exert a great impact on the mechanical properties of joints. The latter is the main contributor to a less reliable connection between aluminum and steel. Iron has a rather low solubility in solid aluminum. When cooled at room temperature, Al-rich intermetallic compounds like Fe_4Al_{13},

$FeAl_3$, Fe_2Al_5 and $FeAl_2$ will be generated even with a small amount of iron. With the increasing content of iron, FeAl, Fe_3Al and other Fe-rich compounds will also be produced. These intermetallic compounds weaken the welded joint and reduce machinability. Therefore, it is of great importance to effectively suppress the generation of Fe-Al intermetallic compounds. Studies showed that adding trace Zr elements to the Zn-15Al welding matrix can reduce the thickness of the generated Fe_4Al_{13} intermetallic compounds and thus decrease the growth rate [7].

By joining method of brazing, the base metal is wetted with liquid filler metal and filling the joint gap. The working temperature is lower than the melting point of the base metal. Therefore, with brazing connection between aluminum and steel, the formation of intermetallic compounds can be effectively suppressed. In the case of brazing aluminum/steel heterogeneous metals, the use of flux is necessary except for vacuum brazing. At present, it is a general agreement that the main benefits of flux in the brazing process include removing oxide film and oil on the base metal surface and preventing further oxidation of the solder by covering the molten liquid solder so as to improve the spreadability and wettability of the brazing material. Currently, low eutectic Nocolock fluxes including KF-AlF_3 and CsF-AlF_3 [8] are widely used to braze aluminum alloys worldwide. This kind of flux has strong activity and can effectively remove the oxide film on the surface of aluminum alloy and promote the wetting and spreading of filler metal on the surface of aluminum and obtain the aluminum–aluminum brazing joint with excellent performance. However, it suffers from poor activity to iron oxide and difficulties in effective removal of the iron oxide film on the surface of carbon steel. Therefore, the liquid filler metal shows less favorable spreadability and wettability to the carbon steel base metal and it is difficult for the filler metal to produce a fine joint between the aluminum and carbon steel. It was confirmed that [9] by adding a small amount of RbF, the activity of CsF-AlF_3 flux on the surface of stainless steel, the oxide film can be improved significantly. With the addition of 0.5 wt.% RbF into CsF-AlF_3 flux, the brazing joint of 6063 aluminum alloy and 304 stainless steel exhibits a tensile strength of up to 127 MPa. Adding 0.5 wt.% Zr [10,11] into Zn-15Al filler metal has an obvious fining effect on the η-Zn phase of the matrix. When the mass fraction of Zr is up to 0.2%, the optimal effect is achieved while the shear strength of brazing joint is up to 143 MPa.

A previous investigation [12] proved that the addition of Ga_2O_3 nanoparticles could significantly improve the performance of CsF-RbF-AlF_3 flux while brazing aluminum alloy to carbon steels. The strengthening mechanism can be mainly attributed to the production of Ga element and its enrichment at the base metal/filler metal interface during brazing. In this paper, the influence of doping trace GaF_3 and Ga_2O_3 nanoparticles into CsF-AlF_3 flux on the spreadability and wettability of Zn-15Al filler metal on the base metals was studied, and enhancement effects of these two nanoparticles on the flux activity were discussed and compared.

2. Materials and Methods

The base metals used in this work were the 6061 aluminum alloy and the Q235 steel, and the compositions of these two materials are respectively listed in Tables 1 and 2. The filler metal was Zn–15Al alloy. A series of CsF–AlF_3–GaF_3 fluxes with different contents of GaF_3 were prepared by using 99.9% GaF_3 flux and GaF_3 of AR purity, and the variation range of GaF_3 is 0.0001–0.1 wt.%. Nano Ga_2O_3 powder in the range of 0.0001–0.1 wt.% were added into the commercial CsF-AlF_3 flux to obtain a series of CsF–AlF_3–Ga_2O_3 fluxes. As an advance preparation, all the filler metal alloys were extruded as a wire with a 2 mm diameter. The base metals for the spreading test were processed into plates with the dimensions of 40 mm × 40 mm × 3 mm in advance. All the specimens and filler metals were mechanically polished using SiC paper. In addition, the above-mentioned experimental materials were degreased with acetone and cleaned using ethanol before brazing.

Table 1. Chemical composition of 6061 aluminum alloy (wt.%).

Alloy	Mg	Si	Cu	Cr	Mn	Zn	Al
6061	1.10	0.61	0.25	0.12	0.01	0.01	Bal.

Table 2. Chemical composition of Q235 steel (wt.%).

Alloy	C	Mn	Si	S	P	Fe
Q235	0.18	0.48	0.30	0.04	0.04	Bal.

The spreading test was carried out strictly in accordance with China's National Standard GB 11364-2008. In this process, 0.1 g filler metal was accurately weighted and placed on the Q235 steel or 6061 aluminum alloy covered with the prepared $CsF–AlF_3–GaF_3$ flux or $CsF–AlF_3–Ga_2O_3$ flux and then carefully put into an electrical resistance furnace as Figure 1. The heating temperature was set at 530 °C and the holding time was 1 min. After the test, the flux residues were collected from the surface of the base metal with clean tweezers and the components in the residues were analyzed with an XRD diffractometer (Brucker D8). CuKα radiation was used in the test, with the voltage of 40 kV, the current of 110 mA, and the scanning speed of 1 °/min. The brazing alloy sample should be polished before test. A 3 mm × 3 mm sample was cut on one side of the brazed joint and used for XRD analysis after cleaning. Then, ultrasonic wave waits were incorporated to clean the test board of base material and calculate the spreading area. In order to ensure the accuracy of the test results, the above tests were repeated 5 times under the same conditions, and the results were averaged.

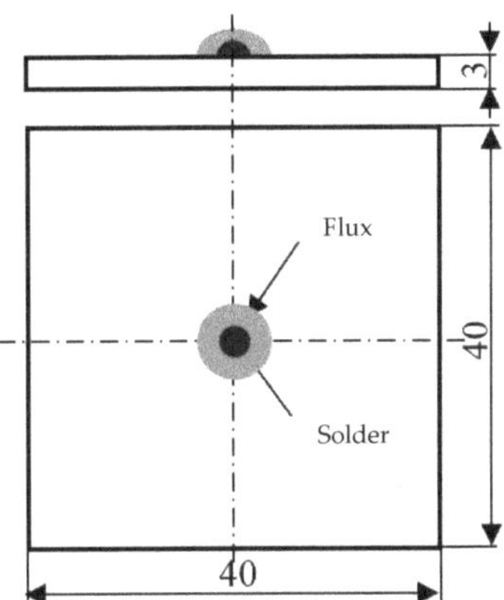

Figure 1. Schematic diagram of the spreading test.

3. Results and Discussion

3.1. The Spreadability and Wettability of Zn-15Al Filler Metal

The effect of GaF_3 and Ga_2O_3 particles on the spreadability of Zn-15Al filler metal was evaluated by the spreading tests that were carried out on the surface of 6061 aluminum alloy and Q235 low-carbon steel with aid of the prepared flux bearing nanoparticles. The spreading areas were measured and averaged. Figure 2a showed the relationship between the concentration of GaF_3 and Ga_2O_3 and spreading area on 6061 aluminum alloy. When the concentration of GaF_3 was 0.01 wt.%, the maximum spreading area of Zn-15Al on 6061 aluminum alloy was 328 mm^2, which is 64% higher than that without GaF_3 addition (200 mm^2). The maximum spreading area of Zn-15Al on 6061 aluminum alloy was 319 mm^2 and was increased by 60% compared with that without doping Ga_2O_3 into the flux, while the concentration of Ga_2O_3 was 0.009 wt.%.

Figure 2b indicates that the spreadability of Zn-15Al filler metal over Q235 steel than Ga_2O_3 was clearly improved by the addition of GaF_3 into the CsF-AlF_3 flux. When the concentration of GaF_3 was 0.0075 wt.%, the maximum spreading area of Zn-15Al on 6061 aluminum alloy was 188 mm^2. When the concentration of Ga_2O_3 was 0.01 wt.%, the maximum spreading area of Zn-15Al on 6061 aluminum alloy was 160 mm^2. Under this circumstance, by adding GaF_3, the spreading area of Zn-15Al filler metal over Q235 steel was increased by 104% comparing to that of CsF-AlF_3 flux (92 mm^2) and by 17.5% comparing to the addition of Ga_2O_3.

Considering the spreading of Zn-15Al filler metal on both 6061 aluminum alloy and Q235 low-carbon steel, the optimum ranges for GaF_3 and Ga_2O_3 in CsF–AlF_3 flux were 0.0075–0.01 wt.% and 0.009–0.01 wt.%, respectively.

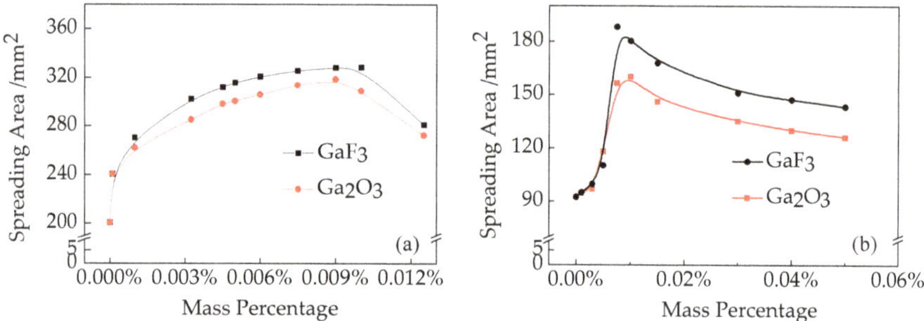

Figure 2. Spreading area of Zn-15Al filler metal (**a**) 6061 aluminum alloy; (**b**) Q235 low-carbon steel.

3.2. The Mechanism of Interfacial Reaction

The CsF-AlF_3-GaF_3 flux could effectively improve the spreadability of Zn-15Al filler metal over 6061 aluminum alloy and Q235 steel, but how it worked was not clear yet and deserved a detailed investigation. Therefore, the theory of interfacial tension was applied reasonably to confirm some potential explanations. On the basis of the Young's equation illustrated by Figure 3 and Equation (1), the liquid balance over solid surface was determined by the interfacial tensions between solids, liquids and gases. Specifically, these interfacial tensions could be summed up as between base metal and molten flux (γ_{SF}), molten filler metal and molten flux (γ_{LF}) and base metal and molten filler metal (γ_{SL}) in this study.

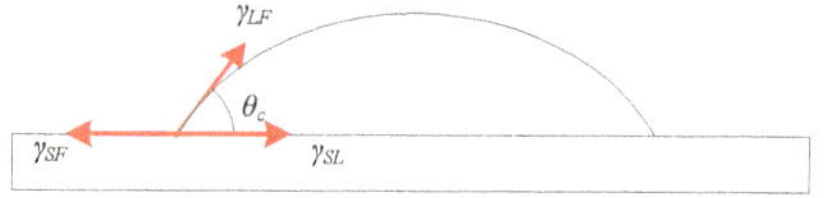

Figure 3. Diagrammatic sketch of Young's equation.

Non-reactive wetting and reactive wetting were two approaches to wetting a solid with molten metal, and the interfacial tension between the base metal and molten filler metal (γ_{SL}) decreased due to the reaction and the spreading was improved. Because the molted flux reacted with the surface metallic oxides over the base metal and removed them, the reactive wetting was what we discussed in this research. While balancing, the relationship of interfacial tensions is shown as Equation (1).

$$\gamma_{SF} = \gamma_{LF} \times \cos\theta + \gamma_{SL} \quad (1)$$

$$\gamma_{SF} > \gamma_{LF} \times \cos\theta + \gamma'_{SL} \quad (2)$$

Alumina reacted with melted XF-AlF_3 (X = Na, K, Rb, Cs) flux and formed AlF_3 to be dissolved, and the interfacial tension of solid–liquid (γ_{SL}) changed from Zn-Al alloy-alumina to

Zn-Al alloy-aluminum [13]. The remarkable intersolubility of Zn and Al made the interfacial tension between AA6061 and Zn-15Al alloy (γ_{SL}) decrease rapidly and the spreading area of Zn-15Al alloy over AA6061 increased significantly under this condition. However, no research has reported that the CsF-AlF_3-based flux could react with iron oxide immediately and the surface oxide was still the largest obstacle. The phase diagram of Fe-Zn showed that intermetallics, such as FeZn and $FeZn_4$, formed below 600 °C. It could be included that the Zn-15Al filler metal spread over Q235 steel made γ'_{SL} decreased gradually by the reaction of Zn and Fe.

The wetting angles of different fluxes over AA6061 are shown in Table 3. Molten Zn-Al filler metal wetted AA6061 alloy and Q235 steel and spread on the base metal, which reduced the γ_{SL} shown in Equation (2) and the balance of Young's equation was broken. The new flux breaks the tension balance of the original interface and thus the spreading area of Zn-15Al is promoted with the addition of GaF_3. Ga^{3+} ion in GaF_3 has a "skin effect", which is analogous to the chemical representation of a thin layer of electrical current concentrated on the outer surface of a conductor, and its small surface tension can significantly reduce γ_{SL}. Due to the enrichment of Ga^{3+} ions, γ_{SL} decreases rapidly, while the spreading area of the liquid filler metal increases further. However, with the consumption of liquid flux, the ability to remove the Al_2O_3 oxide film gradually weakens and finally the reactivity disappears. The interfacial tension (γ_{SF}) between the base metal and flux decreases. The equilibrium of interfacial tension in Equation (1) is reassumed. The spreading area of filler metal attains saturation and the spreading shape is fixed.

Table 3. Wetting angles of different fluxes.

Flux	CsF-AlF_3	CsF-AlF_3-Ga_2O_3	CsF-AlF_3-GaF_3
Wetting angle θ	27.44	19.43	17.71
cosθ	0.89	0.94	0.95
Spreading area/mm^2	200	319	328
Wetting coefficient W	178.0	299.9	311.6

It could be demonstrated that Ga^{3+} was released from GaF_3, whose reaction equation was Equation (3), considering the activity of molten Zn-15Al filler metal. Ga could permeate into the molten filler metal spontaneously with its similar chemical property to Al. The EDX (Energy Dispersive X-ray Spectroscopy) results are shown in Figure 4, and it can be seen that in the interface of base metal and filler metal, a weak peak of Ga appeared. Original balance was broken and the spreading area increased significantly because of the decrease in γ_{SL}. The diagrammatic sketch of the interfacial tension of spread Zn-Al filler metal is shown as Figure 5.

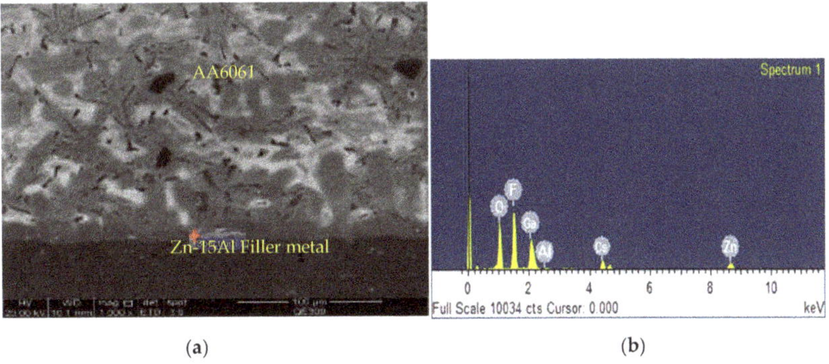

(a) (b)

Figure 4. EDX results of the interface of AA6061 and Zn-15Al filler metal: (**a**) SEM scanning image; (**b**) EDX results of the section.

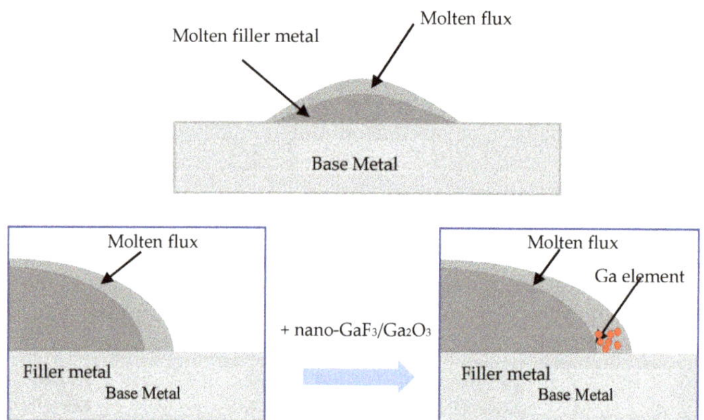

Figure 5. Diagrammatic sketch of the interfacial tension of spread Zn-Al filler metal.

The XRD results of the flux residue were shown in Figure 6. It revealed the residue products over AA6061 were $MgMnSi_2O_6$, Al_2SiO_5, $MgSiO_3$, Zn_2SiO_4, CuO, Cu_4O_3, Al_2O_3, $Cs_{11}O_3$, AlF_3, MnF_2 and $ZnGa_2O_4$, and which were Al_2SiO_5, Fe_2SiO_4, $AlPO_4$, $FePO_4$, Al_2O_3, ZnO, SiO_2, AlF_3, MnF_2 and $MnAlF_5$ over Q235 steel.

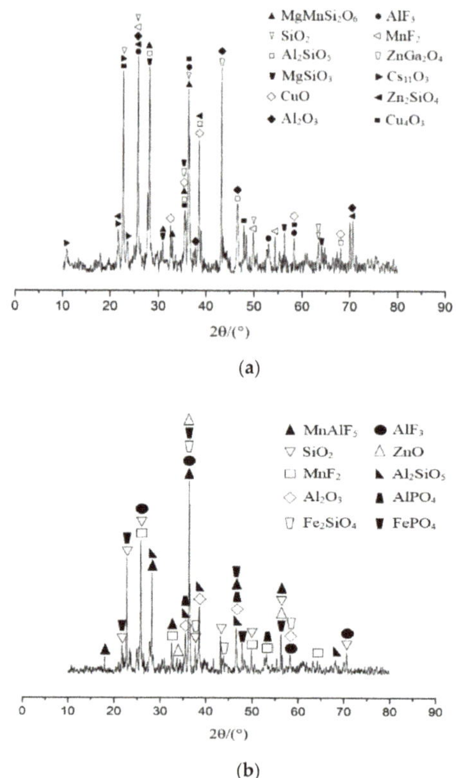

Figure 6. XRD analysis results of flux residua: (**a**) 6061 aluminum alloy; (**b**) Q235 low-carbon steel.

The compounds over AA6061 surface were mainly made up of Mg_2Si, MgO, $MgAl_2O_4$, amorphous Al_2O_3 and a little Cu, Mn, Cr, Fe according to report [14]. The content of Mg, Mn and Si increased while over 520 °C. Figure 6 reveals that the $Cs_{11}O_3$, silicate and fluoride came from the reaction between the oxide and flux. It could be confirmed that $ZnGa_2O_4$ arose from the residue of the trace amounts of GaF_3 in the flux. It was reported [15] the additional Ga tended to enrich on the surface, which led to the speculation that GaF_3 might be involved in some kind of reaction. Because of this, $ZnGa_2O_4$ was validated to show its diffraction peaks in the XRD pattern, whose reaction equation is Equation (4).

$$GaF_3 \rightarrow Ga^{3+} + 3F^- \tag{3}$$

$$4ZnO + 2GaF_3 \rightarrow ZnGa_2O_4 + 3ZnF_2 \tag{4}$$

$$ZnO + Ga_2O_3 \rightarrow ZnGa_2O_4 \tag{5}$$

$$CsFAlF_4 \overset{\Delta}{\leftrightarrow} CsF + AlF_3 \tag{6}$$

$$F^- + Al_2O_3 \rightarrow AlF_3 + O^{2-} \tag{7}$$

$$F^- + Fe_2O_3 \rightarrow FeF_3 + O^{2-} \tag{8}$$

The mechanism for alumina removal from CsF-AlF_3 flux was the dissolution and reaction of the active ingredients such as F^-, HF, $SiF_6{}^{2-}$ and Zn^{2+} ions. Zn^{2+} and $SiF_6{}^{2-}$ ions were liable to be generated during the spreading test because of the massive amount of Zn in the filler metal and a little Si on the base metal surface. The major flux reaction is listed in Equation (5), which refers to the flux with the addition of Ga_2O_3 nanoparticles. In addition, trace amounts of H_2O, NH_4F and NH_4AlF_4 appeared in the CsF-AlF_3-GaF_3 flux, which produced HF and further promoted the removal.

According to Figure 6, the XRD result of residue over Q235 steel was not the same as that in AA6061. Research showed that Fe_2O_3, Fe_3O_4 and FeO were distributed systematically on Q235 steel [16]. Thus, the removal of iron oxide was the primary step. According to the XRD result shown in Figure 6, the reaction of SiO_2, iron oxide and alumina formed Fe_2SiO_5 and Al_2SiO_5. MnF_2 and $MnAlF_5$ were formed by the reaction between F^-, HF, Mn compounds and AlF_3. However, the appearance of a trifle phosphate, $FePO_4$ and $AlPO_4$ in the XRD result led to the conclusion that $FePO_4$ and $AlPO_4$ were formed as Equation (9), under the condition that no phosphorus was contained in the flux and its content was below 0.04 wt.% in Q235 steel.

$$Me_2O_3 + P_2O_5 \rightarrow 2MePO_4(Me = Fe,\ Al) \tag{9}$$

$$Fe_2O_3 + 2Al \rightarrow Al_2O_3 + 2Fe \tag{10}$$

Because the content of phosphorus in Q235 could not be detected by the XRD, it meant that P enriched over Q235 surface, which was a normal phenomenon occurring when heating P-containing alloys such that the phosphorus was burnt out, producing phosphorus oxide. Although the existence of P removed a little of the metallic oxide, it was insufficient to remove all oxide films. Therefore, it was clear that the main reaction mechanism of removing metallic oxides referred to the reaction between Al and iron oxide as Equation (10).

Because of the large concentration of Al and Zn elements in the Zn-Al filler metal, the surface oxide film of the filler metal and Q235 steel is prone to react and the oxide film can be effectively removed. Fe, Cr, Ni and other elements obtained by the reaction were then rapidly dissolved into Q235 steel base material, so Al_2O_3 and ZnO were detected in large quantities in the reaction residue, which reasonably explained the XRD results in Figure 6.

$$36CsF + 2SiO_2 + 6HF + 5MnO \rightarrow 2Cs_{11}O_3 + 2Cs_2SiF_6 + 5Cs_2MnF_6 + 3H_2O \uparrow \tag{11}$$

The reaction shown in Equation (11) explains the decrease in the flux activity due to the huge consumption of the CsF and its lowered flowability. The formation of Ga^{3+} enhanced the spreadability of Zn-15Al filler metal while leading to less activity.

According to the proposed mechanisms, molten flux reacted with oxide and the spread of filler metal was promoted simultaneously. Nevertheless, the efficiency of alumina removal was decreased by the rapid effective loss of molten flux and the spreading speed of Zn-15Al was further delayed. The reason could be the consumption of active substances such as SiF_6^{2-}, HF and F^-, and the production of silicates.

Different from the lower efficiency of CsF-AlF_3 flux on iron deoxidation, the addition of GaF_3 significantly improved the activity on account of the enrichment and production of Ga^{3+}, and thus decreased the interfacial tension between base metal and Zn-15Al [17]. In addition, the dissolving reaction between GaF_3 with ZnO occurred and consequently the spreading of molten Zn-15Al was further promoted.

Instead of reacting and producing Ga^{3+} for the addition of GaF_3, Ga_2O_3 tends to react with ZnO to produce $ZnGa_2O_4$ according to Equation (5). Meanwhile, when nano-Ga_2O_3 particles were added into the flux as a chemical agent, Ga element can only be released by the reaction of Zn-Al alloy and Ga_2O_3 particles. As a typical spinel oxide, $ZnGa_2O_4$ belongs to a cubic crystal system and the chemical stability and thermal stability are very high [18]. Accordingly, more energy is needed to separate Ga from Ga_2O_3. The formation of $ZnGa_2O_4$ somehow facilitates the release of active substances, such as F and Al, which remove the oxide film on the surface of the 6061 Al alloy and the Q235 steel.

Combining the above issues with the XRD results shown in Figure 6, it was speculated that the main reactions are shown as Equations (3)–(11). The formation of active substances such as SiF_6^{2-}, HF and F^- helped the molten flux remove the surface oxide over AA6061, and the iron oxide over Q235 steel was removed by active Al in the filler metal and enriched element P, which led to the improvement in the spreadability of molten Zn-15Al on the base metal.

3.3. Comparative Analysis on the Effect of GaF_3 and Ga_2O_3 Addition

Combined with the analysis of the enhancement mechanism of doping Ga_2O_3 nanoparticles into the flux in a previous study [12], it can be concluded that surface-active Ga obtained from the reaction of the flux and base metals during brazing plays a major enhancement role in promoting the spreading of Zn-Al filler metal on the base metal.

As an ionic compound, GaF_3 consists of cations Ga^{3+} and anions F^-, which means Ga^{3+} and F^- ions can be dissociated more easily from GaF_3 than Ga_2O_3 when heated. The dissociation product Ga^{3+} can rapidly and effectively reduce the interfacial tension between the Zn-15Al filler metal and the base metal to further promote the spreadability of the filler metal. In addition, the synergistic effect of GaF_3 can promote the surface enrichment of P element according to the above "skin effect" and partially remove the surface oxide film of Q235 carbon steel, which further improve the spreadability of the filler metal. Meanwhile when nano-Ga_2O_3 particles were added into the flux as a chemical agent, Ga can only be released by the reaction of Zn-Al alloy and Ga_2O_3 particles. Therefore, compared with GaF_3, more energy is needed to separate Ga from Ga_2O_3 during brazing.

Compared with the results of previous analysis [12,19], the addition of GaF_3 and Ga_2O_3 in this study increases the spreading area of the filler metal on both 6061 aluminum alloy and Q235 steel, and the flux bearing GaF_3 has better activity than that bearing Ga_2O_3.

4. Conclusions

In this study, trace amounts of GaF_3 and Ga_2O_3 nanoparticles were respectively added into CsF-AlF_3 flux for brazing aluminum to carbon steels. The effects of these two nanoparticles on the flux activity and the spreadability of Zn-15Al filler metal on the base metal were comparatively investigated. The major conclusions were as follows:

(1) The spreading tests showed that with the addition of trace amounts of GaF_3 and Ga_2O_3 in CsF-AlF_3 flux, the spreadability of Zn-15Al filler metal both on Q235 steel and AA6061 alloy was effectively improved, and the enhancement effect of GaF_3 was more obvious. The optimal contents of GaF_3 and Ga_2O_3 were 0.0075–0.01 wt.% and 0.009–0.01 wt.%, respectively.
(2) The "skin effect" of GaF_3 improved the activity of fluxes by reacting with Al atoms to produce Ga^{3+}, and enrichment of Ga^{3+} on molten filler metal decreased the interfacial tension and enlarged the spreading area of molten Zn-15Al filler metal on the base metals.
(3) In addition, as an ionic compound, GaF_3 could easily react and generate Ga^{3+} to promote the enrichment of P element in the base metal to remove the oxide film on Q235 carbon steel and promote the wetting of the filler metal. Therefore, the activity of CsF-AlF_3 flux with the addition of GaF_3 was higher than that with the addition of Ga_2O_3.

Author Contributions: Conceptualization, S.X.; methodology, J.Z.; software, Z.Y.; validation, Z.Y.; formal analysis, Z.Y.; investigation, Z.Y.; resources, S.X.; data curation, Z.Y.; writing—original draft preparation, Z.Y.; writing—review and editing, Z.Y.; visualization, Z.Y.; supervision, S.X.; project administration, S.X.; funding acquisition, S.X. All authors have read and agreed to the published version of the manuscript.

Funding: This research was funded by the National Natural Science Foundation of China, Grant No. 51375233 and the Priority Academic Program Development of Jiangsu Higher Education Institutions (PAPD).

Acknowledgments: We gratefully acknowledge the support of Jiangsu Higher Education Institutions for providing a scholarship for Zhen Yao. Songbai Xue acknowledges financial support from the National Natural Science Foundation of China, Grant No. 51375233.

Conflicts of Interest: The authors declare no conflict of interest.

References

1. Choi, C.Y.; Kim, D.C.; Nam, D.G. A Hybrid Joining Technology for Aluminum/Zinc Coated Steels in Vehicles. *Mater. Sci. Technol.* **2010**, *26*, 858–964. [CrossRef]
2. Gale, W.F.; Butts, D.A. Transient liquid phase bonding. *Sci. Technol. Weld. Join.* **2004**, *9*, 283–300. [CrossRef]
3. Yu, S.; Ling, S.; Jiankang, H.; Yufen, G. Performance of $Fe_2A_5Zn_{0.4}$ in Interface of Aluminum and Galvanized Steel Welding-Brazing and Its Formation. *Rare Met. Mater. Eng.* **2013**, *42*, 432–436.
4. Jing, Y.; Xiaoyan, L.; Li, C.; Shuili, G.; Qiaoyan, L. Microstructure and Properties of Twin Spot Laser Welded Joints of 1420 Al-Li Alloy. *Rare Met. Mater. Eng.* **2011**, *40*, 871–874.
5. Jinglong, Y.; Songbai, X.; Peng, X.; Wei, D.; Zhaoping, L.; Man, Z. Microstructure and mechanical properties of aluminum stainless steel brazed joint with torch brazing. *Trans. China Weld. Inst.* **2015**, *36*, 63–66.
6. Zhen, L.; Xuyou, W.; Aiqing, Y.; Guoliang, Q.; Wei, W.; Shangyang, L. Research on Fusion-Brazing Joining between Aluminum and Steel by Laser-MIG Hybrid Welding. *Rare Met. Mater. Eng.* **2009**, *38*, 229–233.
7. Jinglong, Y.; Songbai, X.; Ducan, P.S. An Impact of Zirconium Doping of Zn-Al Braze on the Aluminum-Stainless Steel Joints Integrity During Aging. *J. Mater. Eng. Perform.* **2017**, *26*, 358–365.
8. Songbai, X.; Ling, Z.; Zongjie, H.; Xiang, H. Reaction mechanism between oxide film on surface of Al-Li alloy and CsF-AlF_3 flux. *Trans. Nonferrous Met. Soc. China* **2008**, *18*, 121–125.
9. Jinglong, Y.; Songbai, X.; Peng, X.; Zhaoping, L.; Wei, D.; Junxiong, Z. Development of novel CsF–RbF–AlF_3 flux for brazing aluminum to stainless steel with Zn–Al filler metal. *Mater. Des.* **2014**, *64*, 110–115.
10. Jinglong, Y.; Songbai, X.; Peng, X.; Zhaoping, L.; Weiming, L.; Guangxing, Z.; Qingke, Z.; Peng, H. Development of Zn–15Al–xZr filler metals for Brazing 6061 aluminum alloy to stainless steel. *Mater. Sci. Eng. A* **2016**, *651*, 425–434.
11. Jinglong, Y.; Songbai, X.; Peng, X.; Zhaoping, L. Effect of zirconium on microstructure and properties of Zn-15Al filler metal. *Trans. China Weld. Inst.* **2016**, *12*, 61–65.
12. Zhen, Y.; Songbai, X.; Jinlong, Y.; Junxiong, Z. Inducing the Effect of a Ga_2O_3 Nano-Particle on the CsF-RbF-AlF_3 Flux for Brazing Aluminum to Carbon Steels. *Crystals* **2020**, *10*, 183.
13. Haidong, Q.; Haiyan, G.; Jun, W.; Baode, S. Recent Researches and Development of Brazing Flux for Aluminum. *Mater. Rev.* **2007**, *12*, 76–79.
14. Junxiong, Z.; Songbai, X.; Peng, X.; Shuang, L. Thermodynamic reaction mechanism of the intermetallic compounds of SnxNdy and GaxNdy in soldered joint of Sn–9Zn–1Ga–0.5Nd. *J. Mater. Sci.* **2015**, *26*, 3064–3068.

15. Hong, Z.; Songbai, X.; Zhong, S. Mechanism of CsF-AlF_3 and KF-AlF_3 fluxes reacting with oxide films of 6063 aluminum alloy. *Trans. China Weld. Inst.* **2009**, *30*, 13–16.
16. Yi, W.; Dun, Z.; Huaiqun, L.; Yongjuan, L.; Baorong, H. Influence of sulphate-reducing bacteria on environmental parameters and marine corrosion behavior of Q235 steel in aerobic conditions. *Electrochim. Acta* **2010**, *55*, 1528–1534.
17. Qiyun, Z.; Sunqi, L.; Dongqi, L.; Liangchun, Y. Effect of trace additive on anti-oxidation of molten SN-Pb eutectics. *Trans. China Weld. Inst.* **1984**, *20*, 296–302.
18. Yanbo, J.; Shujie, J.; Shiyong, C.; Dongbo, W.; Jinzhong, W. Synthesis and Photocatalytic Activity of $ZnGa_2O_4$ Nanocubes. *China J. Lumin.* **2019**, *40*, 49–56.
19. Junxiong, Z.; Songbai, X.; Peng, X.; Jinlong, Y.; Zhaoping, L. Effect of Ga_2O_3 on the Wettability and Spreadability of CsF-RbF-AlF_3 Flux/Zn-Al Filler Metal on Aluminum and Steel. *Rare Met. Mater. Eng.* **2017**, *46*, 1900–1904.

Article

Inducing the Effect of a Ga_2O_3 Nano-Particle on the CsF-RbF-AlF_3 Flux for Brazing Aluminum to Carbon Steels

Zhen Yao, Songbai Xue *, Jinlong Yang and Junxiong Zhang

College of Materials Science and Technology, Nanjing University of Aeronautics and Astronautics, Nanjing 210016, China; yaozhen123@nuaa.edu.cn (Z.Y.); yangjinloong@163.com (J.Y.); zhangjunxiong126@163.com (J.Z.)
* Correspondence: xuesb@nuaa.edu.cn; Tel.: +86-025-8489-6070

Received: 15 January 2020; Accepted: 5 March 2020; Published: 7 March 2020

Abstract: In this study, a Ga_2O_3 nano-particle was added into CsF-RbF-AlF_3 flux to develop a highly active flux for brazing aluminum alloy to steel, and the spreadability and wettability of Zn-Al filler metal that matched the CsF-RbF-AlF_3 flux-doped Ga_2O_3 nano-particle on the steel were investigated. The results showed that the spreadability and wettability of the CsF-RbF-AlF_3 flux-doped Ga_2O_3 nano-particle could be remarkably improved when matching Zn-Al filler metals on both aluminum and low-carbon steel, for which the optimal content is in the range of 0.001–0.003 wt.% of Ga_2O_3. An investigation and analysis on the mechanism of reactions among CsF-RbF-AlF_3-doped Ga_2O_3 nano-particle flux and filler metal or base metals showed that the Ga_2O_3 nano-particle is selectively absorbed by the interface of molten Zn-2Al filler metal and base metal, which released the surface-active element Ga to enrich the molten Zn-2Al filler metal and decreased the interfacial tension, so as to promote the enlargement of its spreading area during the brazing process. It was concluded that adding a trace amount of Ga_2O_3 nano-particle into CsF-RbF-AlF_3 flux is a meaningful way to improve the activity of flux for brazing aluminum to steel compared with adding $ZnCl_2$, which poses the risk of corrosion on aluminum.

Keywords: Ga_2O_3; flux; Zn-Al filler metal; wettability; spreadability

---

1. Introduction

Brazing or welding Al to steel has become a hot subject in recent years because Al–steel composite metal materials have been found to have wide applications in industry. In order to develop brazing technology, the improvement of flux is a key problem that must be solved. CsF-AlF_3 eutectic flux has been widely used in brazing aluminum alloys [1] and exhibits an excellent performance in brazing dissimilar metals matched with Zn-Al filler [2,3], such as brazing Al to Al [4] and Al to Cu [5]. However, CsF-AlF_3 flux is not so good when brazing aluminum to steel, and the activity of removing the oxides on Al and steel simultaneously needs to be improved.

It was reported that matching CsF-RbF-AlF_3 flux and the 6061 aluminum alloy to 304 stainless steel can be executed [6]. According to Ref. [7], adding 6 mol.% of $ZnCl_2$ enhanced the activity of CsF-AlF_3 when removing the alumina during brazing. It is well known that the Cl^- ion is a harmful ion to aluminum and steel, so developing a novel flux without a Cl^- ion as an activator represents meaningful work.

In this paper, by adding a micro-nano Ga_2O_3 particle into an CsF-RbF-AlF_3 flux as an activator, the spreadability and wettability of Zn-Al filler metal on the aluminum, along with steel, were investigated. XDR analysis and the reaction mechanism were also studied. The results would be helpful for brazing AA5052 to Q235 steel, without the risk of Cl- ion corrosion.

2. Materials and Methods

The experimental base metals used for spreading tests were commercially supplied AA5052 and Q235 steel (Changshu Huayin Filler Metals Co., Ltd., Suzhou, China) with dimensions of 50 mm × 50 mm × 3 mm. The filler metal was a 2 mm diameter commercial Zn-2Al alloy wire (Zhejiang Xinrui Welding Materials Co., Ltd., Shengzhou, China). The flux used in this research was prepared using nano Ga_2O_3 powder, RbF (AR purity, Shanghai Jiuyi Chemical Reagent Company, Shanghai, China), and commercial CsF-AlF_3 flux (Zhejiang Xinrui Welding Materials Co., Ltd., Shengzhou, China). A total of 0.5 wt% of RbF and nano Ga_2O_3 powder in the range of 0.0001–0.1 wt.% were added to commercial CsF-AlF_3 flux. In order to obtain homogeneous CsF-RbF-AlF_3−xGa_2O_3 flux, accurately weighted Ga_2O_3 powder, RbF, and CsF-AlF_3 were scattered in deionized water, together with ultrasonic concussion, for an hour, in which the weight deviations of Ga_2O_3 and RbF were controlled to ±1 mg. For experimenting conveniently, all specimens were pretreated: they were degreased with acetone, oxides were removed with NaOH solution for AA5052 and HCl for Q235 steel, and the specimens were then washed with deionized water and dried naturally.

The spreading test was carried out according to China's National Standard GB/T11364-2008. A total of 0.12 g filler metal was placed on base metal, covered with the prepared CsF-RbF-AlF_3-Ga_2O_3 flux, and put into the VULCAN3-130 muffle furnace (Nanjing University of Aeronautics and Astronautics, Nanjing, China) holding at 600 °C for 60 s. To obtain accurate results, each flux was tested five times in the same conditions. The flux residue formed during the test was carefully sampled. Then, the experimental specimens were completely cleaned with hot deionized water and dried naturally. The Ga_2O_3 particle was tested by a Hitachi S-4800 SEM (Ningbo Institute of Materials Technology&Engineering, Ningbo, China), and by using the energy spectrum device (energy dispersion X-ray spectrometry, EDX) attached to the equipment, microstructure observation and joint composition analysis were carried out on the solder and joint samples. The spreading areas of filler metals were measured and calculated by Image Pro-plus software, and the flux residue was analyzed with a Bruker D8 XRD analyzer (Ningbo Institute of Materials Technology&Engineering, Ningbo, China).

3. Results and Discussion

3.1. Morphology and Size of the Ga_2O_3 Micro-Nano Particle

In order to improve the activity of CsF-RbF-AlF_3 flux, the particle size of added Ga_2O_3 has to be small enough. A range from 100 nano-meters to several microns is perfect. The size of the Ga_2O_3 particle was tested by SEM, and the results are shown in Figure 1, in which the size of the particle is below 5 μm. The EDX elemental mapping of Ga_2O_3 particles is shown in Figure 2. It was previously reported [8] that Ga_2O_3 particles presenting as short rods have more importance in the integrality of the crystal structure than those presenting as long cylinders.

(a) (b)

Figure 1. Morphology of Ga_2O_3 particles: (**a**) 10,000 times; (**b**) 20,000 times.

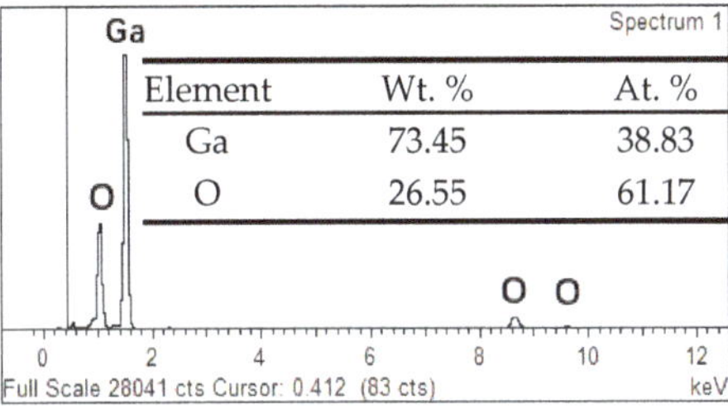

Figure 2. EDX elemental mapping of Ga_2O_3 particles.

3.2. Application Properties of the Zn-2Al Filler Metal

For determining the effect of the Ga_2O_3 particle on the spreadability and wettability of Zn-2Al filler metal, tests were carried out over 5052 aluminum and Q235 low-carbon steel with the new flux, whose spreading areas were measured and averaged. Figure 3 shows the relationship between the Ga_2O_3 concentration and spreading areas, and partial spreading tested graphs are presented in Figure 4.

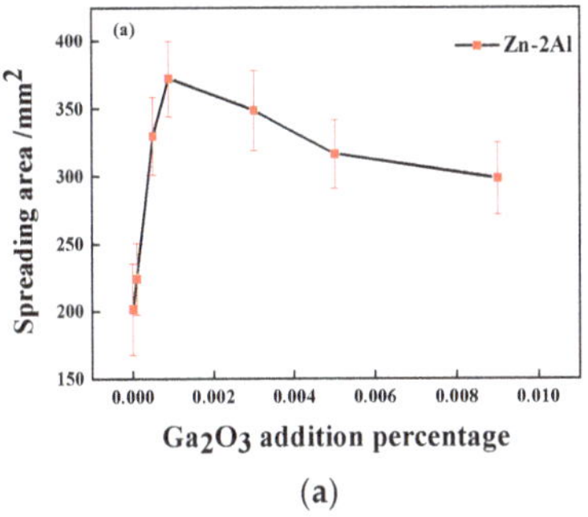

(**a**)

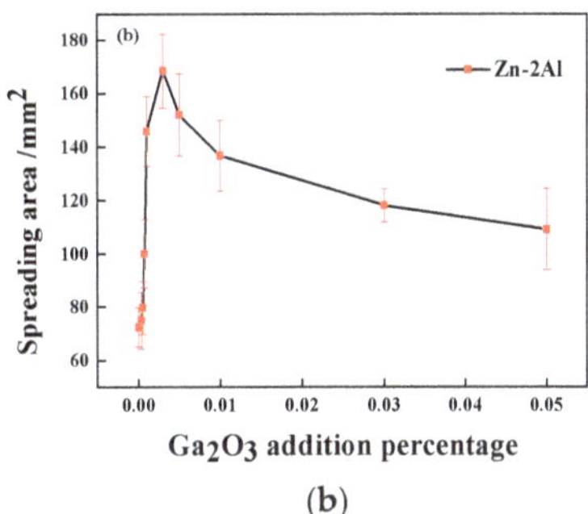

(**b**)

Figure 3. Effect of the Ga_2O_3 concentration on spreading areas of Zn-2Al filler metal on base metal: (**a**) 5052 aluminum; (**b**) Q235 low-carbon steel.

Figure 3 indicates that Ga_2O_3 addition to the CsF-RbF-AlF_3 flux clearly improved the spreadability and wettability of Zn-2Al filler metal on the base metal. It was found that the maximum spreading area of Zn-2Al on 5052 aluminum occurred when the concentration of Ga_2O_3 was 0.0009 wt.%. For Q235 steel, the maximum value was 0.003 wt.%. Under this circumstance, the spreading area of Zn-2Al filler metal was approximately 350 mm^2 over 5052 aluminum, which was an increase of 75% compared to the applications of CsF-RbF-AlF_3 flux. Additionally, it was 160 mm^2 over Q235 steel, representing an improvement of 122%. However, the spreading area mildly decreased with Ga_2O_3 addition continually rising. Based on the above results, the conclusion was that the CsF-RbF-AlF_3-Ga_2O_3 flux could be applied to brazing Q235 steel to AA5052 matched with Zn-2Al filler metal, and the additional concentration of Ga_2O_3 particle range was 0.0009–0.003 wt.%. In Figure 4, partial spreading images are shown. It was obvious that the spreading area of Zn-2Al filler metal was better with the Ga_2O_3 particle added in CsF-RbF-AlF_3 flux.

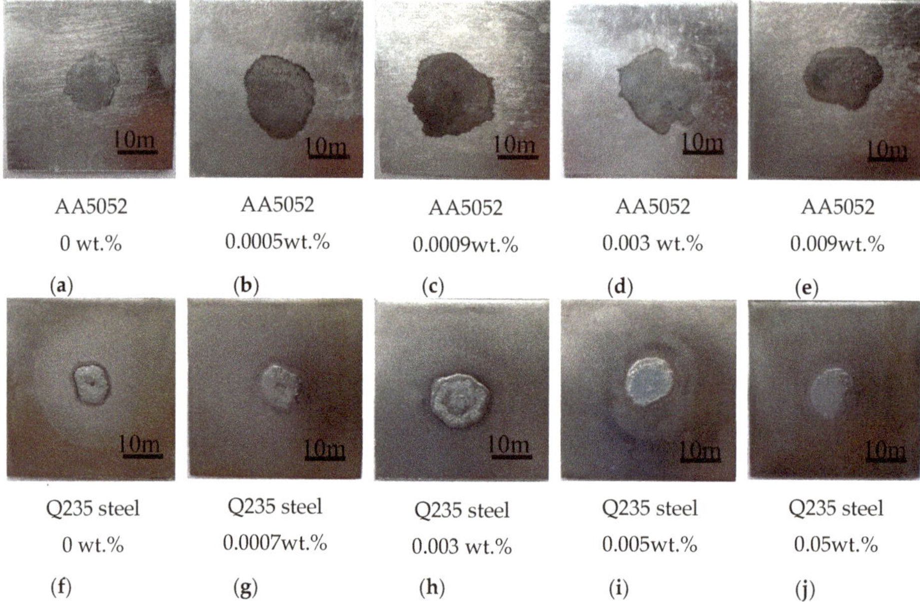

Figure 4. Spreading tests of Zn-2Al filler metal on base metals with CsF-RbF-AlF_3-xGa_2O_3 flux: (**a–e**) 5052 aluminum; (**f–i**) Q235 low-carbon steel.

3.3. Interfacial Induction Effect of the Ga_2O_3 Particle

The CsF-RbF-AlF_3-Ga_2O_3 flux improved the spreadability and wettability of Zn-2Al filler metal over 5052 aluminum and Q235 steel obviously, but the reason for this was not clear and still needed to be investigated. Therefore, we attempted to employ the theory of interfacial tension [9] to identify some explanations. According to Young's equation, the liquid balanced over a solid surface can be determined by the relationship between the interfacial tension of the solid and liquid. In this investigation, a balance of these interfacial tensions was shown between the base metal and molten flux ($\sigma_{\text{s-flux}}$), molten filler metal and molten flux ($\sigma_{\text{l-flux}}$), and base metal and molten filler metal ($\sigma_{\text{s-l}}$).

Dezellus [10] mentioned that there were two kinds of ways in which molten metal wets a solid—non-reactive wetting and reactive wetting—and the reaction of a liquid and solid decreases their interfacial tension ($\sigma_{\text{s-l}}$) to improve spreading. Because the molten flux reacted with the surface metallic oxides over the base metal to remove it, the explanation for the interfacial reactions discussed in this research was that of reactive wetting. While balancing, the relationship between interfacial tensions is in accordance with Equation (1):

$$\sigma_{s-flux} = \sigma_{l-flux} \cdot \cos\theta + \sigma_{s-l} \tag{1}$$

According to the investigation of Qian [11], alumina reacted with melted XF-AlF_3 (X = Na, K, Rb, and Cs) flux and formed AlF_3 to be dissolved, and the interfacial tension of solid–liquid ($\sigma_{\text{s-l}}$) changed from Zn-Al alloy-alumina to Zn-Al alloy-aluminum. Meanwhile, owing to the excellent intersolubility of Zn and Al [12], the interfacial tension of the AA5052 and Zn-2Al alloy ($\sigma_{\text{s-l}}$) decreased rapidly. Therefore, the spreading area of the Zn-2Al alloy over AA5052 increased obviously.

However, the surface oxide was the maximum obstacle to the brazing process for Q235 steel. No research has reported that the CsF-AlF_3 series flux reacted with iron oxidation immediately. Therefore, the reason why the Zn-Al alloy spread over Q235 should be due to the reaction between the Zn-Al alloy and oxidation. The thermodynamically calculated results shown in Table 1 indicated that iron oxides reacted with Al to form simple substance Fe and dissolved [13]. The phase diagram of Fe-Zn showed that intermetallics, such as FeZn and $FeZn_4$, formed below 600 °C. Precisely because of the reaction of Zn and Fe [14], the Zn-2Al filler metal spread over Q235 steel, and their interfacial tension ($\sigma'_{s\text{-}l}$) decreased along with it.

Table 1. Reaction Gibbs energy of iron oxidation and Al (T = 873.15 K).

Compounds	Inductions	ΔG/ kJ·mol^{-1}
FeO	Al	−259.98
Fe_2O_3	Al	−811.13
Fe_3O_4	Al	−1041.32
Ga_2O_3	Al	−575.18

Because of the interfacial reaction, molten Zn-Al filler metal wetted the AA5052 alloy and Q235 steel and spread on them, which broke the balance of Young's equation by reducing the $\sigma_{s\text{-}l}$ shown in Equations (1) and (2). When the gravity was larger than the molten flux, as shown in Table 2, the Ga_2O_3 particle settled from molten flux freely to the molten filler metal surface. According to the result of the thermodynamic calculation shown in Table 1, it was found that the Ga_2O_3 particle released Ga atoms by reacting with the Zn-Al alloy, where Ga was surface active, and could enrich the surface of filler metal to decrease its interfacial tension between metals ($\sigma'_{s-l} < \sigma_{s-l}$) [15]:

$$\sigma_{s-flux} < \sigma_{l-flux} \cdot \cos\theta + \sigma'_{s-l} \tag{2}$$

Table 2. Density of the compound.

Compound	Density/g·mL^{-1}
Ga_2O_3	6.44
$CsAlF_4$	3.7

Considering the activity of molten Zn-2Al filler metal, Ga could be released and tested to demonstrate its release from Ga_2O_3. Due to having a similar property to Al [16], Ga permeated in the molten filler metal spontaneously. The microstructure at the interface of the joint is shown in Figure 5. There is no obvious defect in the interface, the filler metal is fully spread out on the surface of the Q235 steel, and the interface is well-combined. The reaction layers at the interface between the braze joint and base metal differed in thickness: the light gray layer near the braze joint and the dark gray layer near the base metal. The thickness of the interfacial IMC layer highlighted in the pictures was calculated to be 2.82 μm. The thickness of the light gray layer was 1.56 μm, and that of the dark gray layer was 1.26 μm. The EDX results in Figure 6 showed that the weak peak of Ga appeared in the interface of base metal and filler metal. With the increase of Ga_2O_3 addition, Ga_2O_3 particles displayed more activity by presenting as short rods. Owing to the decrease in $\sigma_{s\text{-}l}$, as well as the interfacial reactions, the original balance was broken and the spreading area increased obviously. A description of the variation of the interfacial tensions is shown as Figure 7. According to the Fe-Al binary phase diagram, the dark gray layer should be the θ-phase Fe-Al intermetallic compound. The θ-phase is in equilibrium with the α-Al phase in the Fe-Al phase diagram and has a monoclinic crystal structure. The atomic ratio of Al to Fe ranges from 3 to 3.5 in the θ-phase, and the crystal structure is usually represented by Fe_4Al_{13}, $FeAl_{3.2}$, and $FeAl_3$.

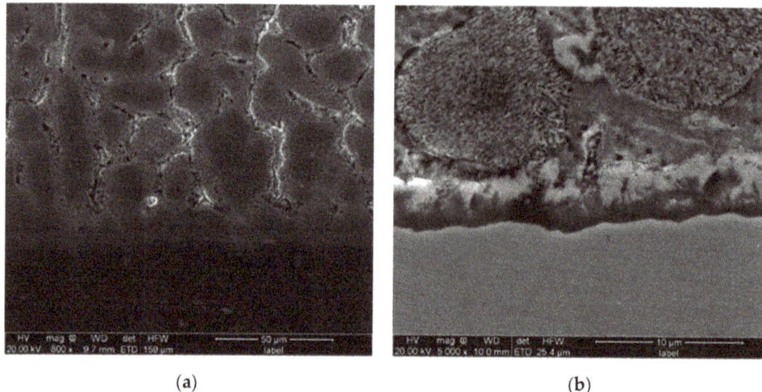

Figure 5. Microstructure at the interface of the joint: (**a**) 5052 aluminum; (**b**) Q235 low-carbon steel.

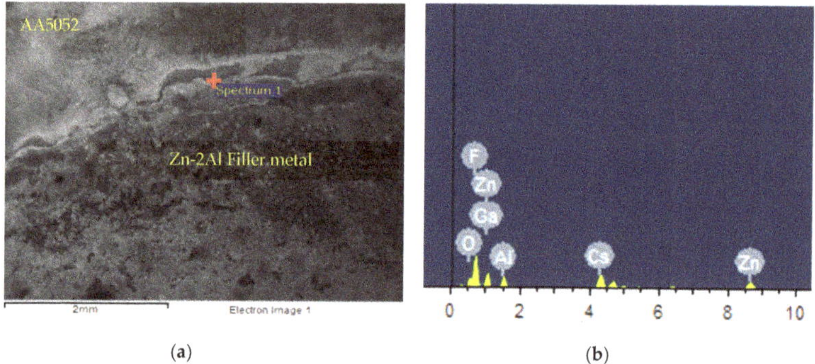

Figure 6. The interface of AA5052 and Zn-2Al filler metal: (**a**) SEM scanning image; (**b**) EDX results of the section.

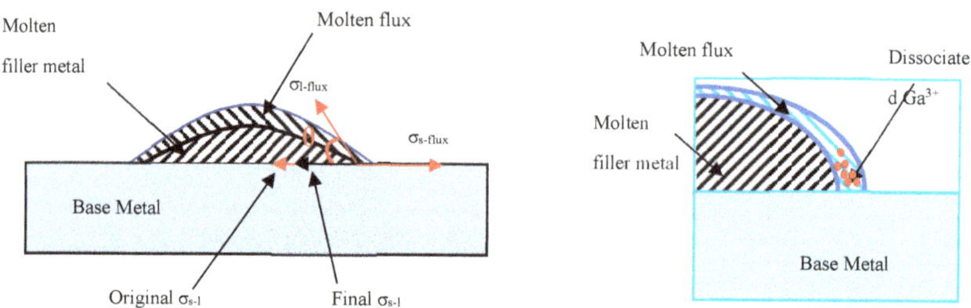

Figure 7. Diagrammatic sketch of Zn-Al spreading on base metal.

3.4. Analysis of the Flux Residue

An XRD analysis experiment of the flux residue collected from the surface of AA5052 and Q235 steel was carried out. It revealed that the residue products over AA5052 were silicate, oxidation, fluoride, $MgMnSi_2O_6$, SiO_2, Al_2SiO_5, $MgSiO_3$, Al_2O_3, AlF_3, MnF_2, $ZnGa_2O_4$, and $Cs_{11}O_3$; however, they were $MnAlF_5$, SiO_2, Al_2O_3, MnF_2, AlF_3, ZnO, Al_2SiO_5, $AlPO_4$, and $FePO_4$ over Q235 steel.

According to a previous report [17], the compounds found over the AA5052 surface were mainly Mg_2Si, MgO, $MgAl_2O_4$, and amorphous Al_2O_3 and a little Cu, Mn, Cr, and Fe. When heating to over 520 °C, Mg, Mn, and Si were enriched. The $Cs_{11}O_3$, silicate, and fluoride were from the reaction between the oxidation over the base metal and CsF-RbF-AlF_3-Ga_2O_3 flux. However, $ZnGa_2O_4$ was identified from the residue, which certainly came from the reaction of trace amounts of Ga_2O_3 in the flux. It was previously reported [18] that additional Ga in the filler metal was enriched on its surface, which meant that Ga_2O_3 might have been enriched. It could be speculated that the reaction equation of $ZnGa_2O_4$ can be written as follows:

$$ZnO + Ga_2O_3 \rightarrow ZnGa_2O_4 \Delta G = -653.15 \text{ KJ/mol } (T = 600\ ^\circ C) \tag{3}$$

It was reported that the mechanism through which $CsF\text{-}AlF_3$ flux removed alumina was a process of dissolution and reaction, and the active ingredients were F^-, HF, SiF_6^{2-}, and Zn^{2+} ions. In this study, a massive Zn element was present in the filler metal, with a small Si element on the base metal surface, which provided the forming conditions for Zn^{2+} and SiF_6^{2-} ions during the spreading test. According to the basic principles of the chemical reaction [19], the major flux reaction was listed in Equation (4) [20]. Because of the imperfections of flux preparation technology, a trace amount of H_2O, NH_4F, and NH_4AlF_4 appeared in the $CsF\text{-}RbF\text{-}AlF_3\text{-}Ga_2O_3$ flux, which released active HF during the experiments and accelerated the removal of alumina:

$$CsAlF_4 \overset{\Delta}{\leftrightarrow} CsF + AlF_3 \tag{4}$$

$$NH_4AlF_4 \overset{\Delta}{\leftrightarrow} NH_4F + AlF_3 \tag{5}$$

$$NH_4F \overset{\Delta}{\rightarrow} NH_3 + HF \tag{6}$$

$$2AlF_3 + 3H_2O \overset{\Delta}{\leftrightarrow} Al_2O_3 + 6HF \uparrow \tag{7}$$

However, the XRD result of the residues over Q235 steel was not the same as that for AA5052, which included $MnAlF_5$, SiO_2, AlF_3, MnF_2, Al_2SiO_5, Al_2O_3, ZnO, $AlPO_4$, and $FePO_4$. Research has shown that Fe_2O_3, Fe_3O_4, and FeO orderly cover Q235 steel [21]. Therefore, iron oxide removal was the first step for brazing conveniently. It could be found that F^- and HF reacted with Mn compounds and AlF_3 to form MnF_2 and $MnAlF_5$. Fe_2SiO_5 and Al_2SiO_5 were obtained from the reaction of SiO_2 and iron oxidation and alumina. Nevertheless, a trifle phosphate, $FePO_4$, and $AlPO_4$ appeared in the XRD result. As it was shown, there was no phosphorus in the flux, whose component in Q235 was below 0.04 wt.% [22]. Therefore, $FePO_4$ and $AlPO_4$ were formed by the reactions between metallic oxides and phosphorus in Q235 steel as

$$2M_2O_3 + 4P + 5O_2 \rightarrow 4MPO_4 (M = Fe, Al). \tag{8}$$

Because the content of phosphorus in Q235 was below the XRD detectability, it meant that P was enriched over the Q235 surface, which is a normal phenomenon that occurs upon heating a P-contained alloy in which phosphorus is lost through burning and phosphorus oxide is formed. Although the existence of P removed a little metallic oxide, it did not remove all oxide films. Therefore, the main reaction mechanism of removing metallic oxides was the reaction between Al and iron oxidations shown in Table 1.

It was concluded that the main reactions between the $CsF\text{-}RbF\text{-}AlF_3\text{-}Ga_2O_3$ flux and surface compounds on the base metal could be described as Equations (3)–(15). Due to the formation of active substances—SiF_6^{2-}, HF, and F^-—the molten flux cleaned up the surface oxidation over AA5052, and active Al in the filler metal and enriched P removed the iron oxidation over Q235 steel. Both of these mechanisms improved the spreadability and wettability of molten Zn-2Al filler metal on the base metal.

However, the reactions between silicate and metallic oxide shown in Equation (9) decreased the flux activity by consuming the CsF in $CsF\text{-}AlF_3$ and inhibited its flowability. Owing to the formation of Ga, the spreadability and wettability of Zn-2Al filler metal were enhanced, despite the activity loss:

$$28CsF + 3SiO_2 + 10HF + MnO \rightarrow 2Cs_{11}O_3 + 2Cs_2SiF_6 + 5MnF_2 + 5H_2O \uparrow \tag{9}$$

$$Mg_2Si + 2O_2 \rightarrow 2MgO + SiO_2 \quad (10)$$

$$MgAl_2O_4 + SiO_2 \rightarrow MgSiO_3 + Al_2O_3 \quad (11)$$

$$Al_2O_3 + SiO_2 \rightarrow Al_2SiO_5 \quad (12)$$

$$MnO + MgO + 2SiO_2 \rightarrow MgMnSi_2O_6 \quad (13)$$

$$Cs_2SiF_6 \xrightarrow{\Delta} 2CsF + SiF_4 \uparrow \quad (14)$$

$$2MnO + 4HF + AlF_3 \rightarrow MnF_2 + MnAlF_5 + 2H_2O \uparrow \quad (15)$$

According to the proposed mechanisms, it was indicated that molten flux reacted with oxidation over base metals to promote the spread of filler metal simultaneously. However, the rapidly effectiveness loss of molten flux decreased the efficiency of alumina removal and further strangled the spreading speed of filler metal, which was mainly due to the consumption of active substances—SiF_6^{2-}, HF, and F^-—and the appearance of silicates.

Compared to the lower efficiency of CsF-RbF-AlF_3 flux on iron oxidation, the Ga_2O_3 addition obviously improved the activity of the flux because of its enrichment and production of Ga, which helped to decrease the interfacial tension between base metal and molten filler metal. Meanwhile, Ga_2O_3 partially reacted with ZnO to dissolve it, in order to promote the further spreading of molten Zn-2Al filler on the base metal.

4. Conclusions

In this study, it was demonstrated that the activity of CsF-RbF-AlF_3 flux was clearly enhanced by adding a trace amount of Ga_2O_3, promoting the further enlargement of the spreading area of the Zn-2Al filler metal over AA5052 and Q235 steel. The major conclusions were as follows.

The spreading tests showed that the addition of trace amounts of Ga_2O_3 in CsF-RbF-AlF_3 flux promoted the enlargement of the spreading area of Zn-2Al filler metal on both Q235 steel and AA5052 matrix, and the optimal content of Ga_2O_3 was 0.001–0.003 wt.%.

The interfacial effect and induction effect of Ga_2O_3 improved the flux activity by reacting with Al atoms to provide Ga and enriched molten filler metal to decrease its interfacial tension and promote the spreading area increase of molten Zn-2Al filler metal over the matrix. Meanwhile, the phosphorus oxidation and active Al removed iron oxidation and effectively increased the spreading area of molten Zn-2Al filler metal over the Q235 steel matrix.

As a result of the identified reaction mechanism, it was found that Ga_2O_3 partially reacted with ZnO and slowed down the efficiency loss speed of CsF-RbF-AlF_3 flux to improve its flowability.

Author Contributions: Conceptualization, S.X.; methodology, J.Z.; software, Z.Y.; validation, Z.Y.; formal analysis, Z.Y.; investigation, Z.Y.; resources, S.X.; data curation, Z.Y.; writing—original draft preparation, Z.Y.; writing—review and editing, Z.Y.; visualization, Z.Y.; supervision, J.Y.; project administration, J.Y.; funding acquisition, S.X. All authors have read and agreed to the published version of the manuscript.

Funding: This work was funded by the National Natural Science Foundation of China, Grant No. 51375233 and the Priority Academic Program Development of Jiangsu Higher Education Institutions (PAPD).

Acknowledgments: We gratefully acknowledge the support of Jiangsu Higher Education Institutions for providing a scholarship for Zhen Yao. Songbai Xue acknowledges financial support from the National Natural Science Foundation of China, Grant No. 51375233.

Conflicts of Interest: The authors declare no conflicts of interest.

References

1. Cheng, F.J.; Yang, J.X.; Zhao, H.W.; Zhang, G.X.; Huang, J.L.; Long, W.M. Relationship between phase structure and melting characteristics of low cesium KF-CsF-AlF_3 aluminum flux. *Trans. China Weld. Inst.* **2013**, *34*, 5–8.

2. Cheng, F.J.; Jin, W.S.; Wang, Y.; Wang, D.P. The Chemical Reactions in Solution Synthesis of Aluminum Alloy Brazing Fluorate Flux. *Adv. Mater. Res.* **2011**, *291*, 924–928. [CrossRef]
3. Xu, Z.W.; Yan, J.C.; Wang, C.; Yang, S.Q. Substrate oxide undermining by a Zn–Al alloy during wetting of alumina reinforced 6061 Al matrix composite. *Mater. Chem. Phys.* **2008**, *112*, 831–837. [CrossRef]
4. Luo, W.; Wang, L.T.; Wang, Q.M.; Gong, H.L.; Yan, M. A new filler metal with low contents of Cu for high strength aluminum alloy brazed joints. *Mater. Des.* **2014**, *63*, 263–269. [CrossRef]
5. Ji, F.; Xue, S.B.; Dai, W. Reliability studies of Cu/Al joints brazed with Zn–Al–Ce filler metals. *Mater. Des.* **2012**, *42*, 156–163.
6. Yang, J.L.; Xue, S.B.; Xue, P.; Lv, Z.P.; Dai, W.; Zhang, J.X. Development of novel CsF–RbF–AlF_3 flux for brazing aluminum to stainless steel with Zn–Al filler metal. *Mater. Des.* **2014**, *64*, 110–115. [CrossRef]
7. Xiao, B.; Wang, D.P.; Cheng, F.J.; Wang, Y. Oxide film on 5052 aluminium alloy: Its structure and removal mechanism by activated CsF–AlF_3 flux in brazing. *Appl. Surf. Sci.* **2015**, *337*, 208–215. [CrossRef]
8. Ma, F.; He, J.H.; Liang, C. Influence of Ga_2O_3 Morphology on Synthesis of YAGG: Ce^{3+} Phosphor Powders by Solid-state Reaction Method. *Guangzhou Chem. Ind. Technol.* **2018**, *19*, 36–38.
9. Weber, C.; Stanjek, H. Energetic and entropic contributions to the work of adhesion in two-component, three-phase solid–liquid–vapour systems. *Coll. Surf. A Physicochem. Eng. Asp.* **2014**, *441*, 331–339. [CrossRef]
10. Dezellus, O.; Eustathopoulos, N. Fundamental issues of reactive wetting by liquid metals. *J. Mater. Sci.* **2010**, *45*, 4256–4264. [CrossRef]
11. Qian, H.; GAO, H.; Sun, B. Recent Researches and Development of Brazing Flux for Aluminum. *Mater. Rev.* **2007**, *12*, 76–78.
12. Osório, W.R.; Garcia, A. Modeling dendritic structure and mechanical properties of Zn–Al alloys as a function of solidification conditions. *Mater. Sci. Eng. A* **2002**, *325*, 103–111. [CrossRef]
13. Fan, R.H.; Lv, H.L.; Sun, K.N.; Wang, W.X.; Yi, X.B. Kinetics of thermite reaction in Al-Fe_2O_3 system. *Thermochim. Acta* **2006**, *440*, 129–131. [CrossRef]
14. Dong, H.; Yang, L.; Dong, C.; Kou, S. Improving arc joining of Al to steel and Al to stainless steel. *Mater. Sci. Eng. A* **2012**, *534*, 424–435. [CrossRef]
15. Zhao, H.T.; Shi, Y.Q.; Tian, M. Investigation on the Effect of the Substrate Temperature on the Photoelectric Properties of the Ga Doped ZnO Film. *Mater. Sci. Forum* **2016**, *852*, 1066–1069. [CrossRef]
16. Guo, S.L.; Li, D.F.; Wu, X.P.; Xu, X.Q.; Du, P.; Hu, J. Characterization of hot deformation behavior of a Zn–10.2Al–2.1Cu alloy using processing maps. *Mater. Des.* **2012**, *41*, 158–166. [CrossRef]
17. Zhang, J.X.; Xue, S.B.; Xue, P.; Liu, S. Thermodynamic reaction mechanism of the intermetallic compounds of Sn_xNd_y and Ga_xNd_y in soldered joint of Sn–9Zn–1Ga–0.5Nd. *J. Mater. Sci. Mater. Electron.* **2015**, *26*, 3064–3068. [CrossRef]
18. Zhu, H.; Xue, S.B.; Sheng, Z. Effect of alloying elements on intermediate temperature filler metal in stepped welding of 6063 aluminum alloy. *Trans. China Weld. Inst.* **2009**, *30*, 33–36.
19. Khan, T.I.; Kabir, M.H.; Bulpett, R. Effect of transient liquid-phase bonding variables on the properties of a micro-duplex stainless steel. *Mater. Sci. Eng. A* **2004**, *372*, 290–295. [CrossRef]
20. Davies, G.B.; Krüger, T.; Coveney, P.V.; Harting, J.; Bresme, F. Interface deformations affect the orientation transition of magnetic ellipsoidal particles adsorbed at fluid-fluid interfaces. *Soft Matter* **2014**, *10*, 6742–6748. [CrossRef]
21. Wan, Y.; Zhang, D.; Liu, H.Q.; Li, Y.J.; Hou, B.R. Influence of sulphate-reducing bacteria on environmental parameters and marine corrosion behavior of Q235 steel in aerobic conditions. *Electrochim. Acta.* **2010**, *55*, 1528–1534. [CrossRef]
22. Wang, X.; Yan, Q.; Li, X. Microstructure and properties of 20 steel pipe joints by transient liquid phase bonding and high temperature brazing. *Trans. China Weld. Inst.* **2005**, *26*, 56–60.

Article

Effect of Bi Substitution on Structural and AC Magnetic Susceptibility Properties of $Nd_{1-x}Bi_xMnO_3$

Nurul Atiqah Azhar [1], Intan Solehah Ismail [1], Nur Baizura Mohamed [2], Azhan Hashim [2] and Zakiah Mohamed [1,*]

1 Faculty of Applied Sciences, Universiti Teknologi MARA, UiTM, Shah Alam 40450, Malaysia; nurulatiqahazhar97@gmail.com (N.A.A.); intansolehah96@gmail.com (I.S.I.)
2 Faculty of Applied Sciences, Universiti Teknologi MARA, UiTM, Pahang 26400, Malaysia; nurbaizura620@uitm.edu.my (N.B.M.); dazhan@uitm.edu.my (A.H.)
* Correspondence: zakiah626@uitm.edu.my; Tel.: +60-183-708-386

Received: 1 June 2020; Accepted: 16 June 2020; Published: 17 June 2020

Abstract: This study synthesizes the neodymium-based manganites with Bi doping, $Nd_{1-x}Bi_xMnO_3$ (x = 0, 0.25 and 0.50) using the solid-state reaction route. The crystal structural, morphological and magnetic properties were determined using X-ray diffraction (XRD), fourier transform infrared spectroscopy (FTIR), field emission scanning electron microscopy (FESEM) and AC magnetic susceptibility. The Rietveld refinement confirmed that the compounds were in the single-phase orthorhombic structure of the $NdMnO_3$ with Pbnm space group and lattice parameter *b* increased with doping from 5.5571 (x = 0) to 5.6787 (x = 0.5). FTIR spectra showed that absorption bands were located within the range of 550–600 cm^{-1}, which corresponded to the Mn–O stretching vibration. FESEM exhibited homogenous compound. The AC magnetic susceptibility measurement studies showed a strong antiferromagnetic (AFM) to paramagnetic (PM) transition existed at 76 K, 77 K and 67 K for samples (x = 0, 0.25 and 0.50), respectively.

Keywords: X-ray diffraction; perovskites manganites; AC magnetic susceptibility

---

1. Introduction

Existing studies on the family of manganites perovskites, $RMnO_3$ as a parent of mixed valence perovskites, where R is made up of rare earth elements, such as La, Pr and Nd, were found a few decades ago due to their exotic behavior in electrical and magnetic properties [1,2]. Manganites perovskites were the favorite amongst researchers compared with other families of oxides because of their various possible substitutions that may change their crystal structure and properties [3]. Since the rediscovery of colossal magnetoresistance (CMR), many researchers have been focusing on the doped compound manganites due to their fascinating basis in physics and significance in technological importance, such as magnetic sensors [4]. Few theories have surfaced to explain the CMR mechanisms, such as double exchange (DE), superexchange (SE), Jahn-Teller (JT) effect, polaronic effects and so on.

Primarily, undoped $NdMnO_3$ is an antiferromagnetic (AFM) insulator with Néel temperature 67.2 K due to the interplay between the ferromagnetic (FM) DE and AFM SE interactions present in the compound [5]. Chatterji et al. confirmed that $NdMnO_3$ has an A-type AFM structure of the Mn magnetic sublattice below Néel temperature, $T_N \approx 81.7$ K [6]. When doped with divalent cations on the Nd sites, for example in $Nd_{1-x}Ca_xMnO_3$, FM semiconducting behavior was observed with $x \leq 0.25$. However, when $x > 0.80$, the compound exhibited AFM semiconducting at low temperatures [7]. For $Nd_{0.5}Ca_{0.5}MnO_3$, the compound undergoes a charge ordering (CO) transition at $T_{CO} = 250$ K and the AFM ordering transition at $T_N = 160$ K, which is consistent with the study done by Wu et al. [7,8]. In this case, both authors suggested that the presence of CO was due to the JT distortion, resulting in the distortion of MnO_6 octahedral and localization of charge carrier, by decreasing its mobility, which seemed to weaken the DE interaction [7–9]. Meanwhile, when $NdMnO_3$ was doped with monovalent cations such as Na^+ and K^+, it showed an FM insulator behavior and CO transition was observed for $Nd_{1-x}Na_xMnO_3$ at −250 K when x > 0.1 [5,10]. However, a lack of knowledge still exists on Nd series doped with trivalent cations such as $Nd_{1-x}Bi_xMnO_3$.

Previous studies reported that although Bi^{3+} (1.24 Å) and La^{3+} (1.22 Å) have almost similar ionic radii, both systems exhibit different structures and properties [11]. $LaMnO_3$ exhibited an AFM insulator [12], while $BiMnO_3$ exhibited an FM insulator with T_C of 105 K–110 K [11,13,14]. Recent studies on bismuth-doped manganites, such as $La_{0.7-x}Bi_xAg_{0.3}MnO_3$ [12], $La_{1-x}Bi_xMnO_3$ [13] and $La_{0.67-x}Bi_xSr_{0.33}MnO_3$ [15], showed a decreased metal–insulator transition temperature, T_{MI}, with an increase of Bi content. The author suggested that the behavior of the latter was related to possible hybridization of Bi $6s^2$ lone pairs and e_g electron of Mn^{3+}, which could have affected the localization of carriers and weakened the DE interaction. Moreover, Bi substitution may also reduce the Mn–O–Mn angle, which can block the movement on e_g electron from Mn^{3+} to Mn^{4+}, which decreases the mobility of conduction electron and weaken the DE process [2,12].

Several works were reported on Bi substitution, especially in La-based system, but a lack of study on Bi substitution in Nd-based system still exists, especially for optical and magnetic properties. In this study, we focus on reporting the structural, optical and magnetic properties of $Nd_{1-x}Bi_xMnO_3$ to investigate the unique role of Bi in the system.

2. Materials and Methods

Polycrystalline samples for $Nd_{1-x}Bi_xMnO_3$ (x = 0, 0.25 and 0.50) were synthesized by using conventional solid state reaction method. High purity (≥99.9%) neodymium oxide (Nd_2O_3), bismuth oxide (Bi_2O_3) and manganese oxide (Mn_2O_3) powders were weighed in stoichiometric ratio using electronic balance and mixed together. Agate mortar and pestle were used to grind the powders for about 2 h before calcination in air at 700 °C for 24 h, repeating twice in order to eliminate volatile foreign particles. Finally, the obtained powders were then pressed into pellet under a pressure of 5 tons and were sintered at 1100 °C for 24 h.

The structure and phase purity of the samples were observed by using the X–ray diffraction technique (X'pert PRO MPD) at room temperature with Cu–K_α (λ = 1.5418 Å) radiation by using a PANanalytical diffractometer. The XRD data were analyzed using the Rietveld refinement method using several software programs, such as General Structure Analysis System (GSAS) and Graphical User Interfaces (EXP–GUI) for the refinement and CMPR to convert data from ASC file to DAT file. Fourier transform infrared spectroscopy (FTIR) results were recorded using FTIR–Raman Drift Nicolet 6700 in the range of 450–2000 cm^{-1} to directly probe with the functional group present in perovskite samples. Samples were thoroughly mixed in KBr before the characterization. LEO model 982 Gemini Field Emission Scanning Electron Microscopy (FESEM) with energy-dispersive spectroscopy (EDX) were used to determine the surface morphology and homogeneity of the perovskite samples. AC susceptibility (χ') measurements were carried out using a Stanford research model SR-7265 lock-in amplifier and were performed in the temperature range from 30 K to 300 K.

3. Results

3.1. Structural Analysis

This study obtained all the samples of $NdMnO_3$ doped with Bi as black and well-crystallized powders. Figure 1 shows the XRD patterns for manganites $Nd_{1-x}Bi_xMnO_3$ (x = 0, 0.25 and 0.50) that were conducted at room temperature to validate the structure and sample purities and were then analyzed by using the Rietveld refinement technique (Figure 2). All samples were indexed in an orthorhombic structure ($a \neq b \neq c$ and $\alpha = \beta = \gamma = 90°$) with Pbnm space group. The refined patterns were quite acceptable as the obtained value of agreement factor, χ^2, showed a value of ~1, which indicated a good fit. Rietveld refinement showed the samples x = 0 and x = 0.25 crystallized in a single phase and were chemically pure. However, a small amount of Bi_2O_3 phase was detected in $Nd_{0.5}Bi_{0.5}MnO_3$ sample. The appearance of a secondary phase and redistribution of peak intensities in the $Nd_{0.5}Bi_{0.5}MnO_3$ sample was an indication of incomplete crystallization of the sample at the sintering temperature. The assumed structure for $NdMnO_3$ was consistent with most previous studies [6,10,16,17] that reported the same structure with Pbna or Pbnm space group (different setting), ICSD No. 153214, 247143. The sharpening of the peaks was due to the better crystallinity of the nanoparticles and only minimal impurities were detected in the XRD pattern of the x = 0.50 sample. In addition, the main diffraction peaks corresponded to (121) hkl plane and matched with the previous study [18]. The crystallite size (D) of the samples estimated by the Debye–Scherer equations, $D = K\lambda/(\beta(\theta)\cos\theta)$, where D is the crystallite size (nm), K is constant with 0.94, λ is wavelength of XRD that is 0.1541 nm for CuK_α radiation Å and β is full-width at half-maximum (FWHM) and θ is the angle of peak of XRD. The average crystallite size was found to be in the range 74–128 nm.

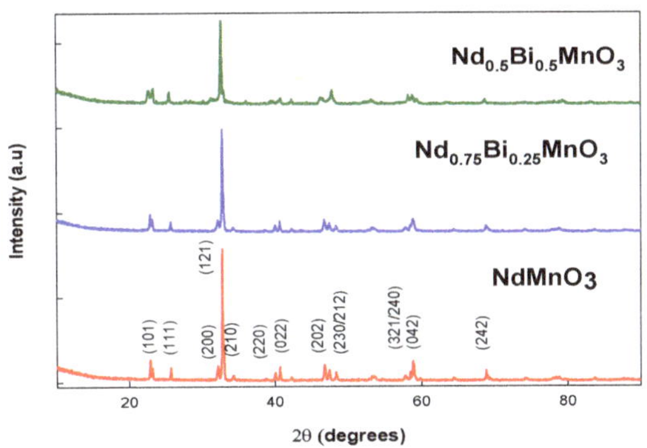

Figure 1. X-ray diffraction of $Nd_{1-x}Bi_xMnO_3$ (x = 0, 0.25 and 0.50) samples.

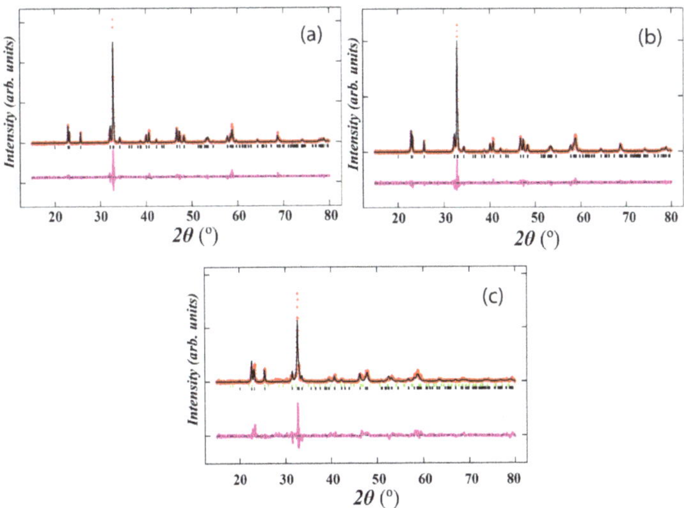

Figure 2. Rietveld refinement for (**a**) $NdMnO_3$, (**b**) $Nd_{0.75}Bi_{0.25}MnO_3$ and (**c**) $Nd_{0.5}Bi_{0.5}MnO_3$. Black solid lines are observed data, the solid red line is the calculated patterns and the pink line is the difference. Tick marks indicate the allowed Bragg reflections. Bi_2O_3 impurity phase was detected for $Nd_{0.5}Bi_{0.5}MnO_3$ and indicated by green tick marks.

Table 1 presents the corresponding data of lattice parameters obtained from Rietveld refinement technique. For the undoped sample, the lattice parameters obtained, a = 5.424 Å, b = 5.5570 Å and c = 7.6648 Å, were found to be quite consistent with the value reported by the previous study [10]. The structural parameter b of the samples increased as the Bi doping increased up to x = 0.50. However, for both lattice parameters, a and c showed no systematic trend due to Bi substitution. The unit cell volume of the samples also increased with the increase of the Bi^{3+} substitution. The increases in unit cell volume were suggested due to the Bi substitution at Nd site, where the Bi^{3+} had a slightly larger ionic radius compared with Nd^{3+} and could replace it, which led to the expansion of cell volume [12,13]. Both authors stated that the radius of Bi^{3+} may depend on the character of $6s^2$ lone pairs; it can be seen that when Bi $6s^2$ lone pair is dominant, the ionic radius of Bi^{3+} is 1.24 Å, while when not, the ionic radius of Bi^{3+} is approximately 1.16 Å [12]. Hence, the increase in structural parameter, b, and cell volume, V, with Bi substitution suggests that the Bi $6s^2$ lone pair is dominant in the $Nd_{1-x}Bi_xMnO_3$ system. From Table 2, the Mn–O bond length decreased and the average ionic radius, r_A, increased as Bi concentration level increased. However, Mn–O–Mn bond angle showed no trend with the increased Bi-doped. The tolerance factor (τ) was calculated using formula $\tau = (<r_A> + <r_O>)/\sqrt{2}\ ((<r_B> + <r_O>)$, where $<r_A>$, $<r_B>$ and $<r_O>$ represent the average ionic radii of A, B and O sites, respectively [9]. The values of τ showed an increase from 0.515 (x = 0) to 0.529 (x = 0.50), as seen in Table 2. The increase in τ indicates that a slight reduction of MnO_6 distortion exists, thus the lattice reduced the mismatch between A–O and B–O layer as the Bi concentration increased [18,19].

Table 1. Lattice parameters *a*, *b* and *c*, and cell volume, *V*, of $Nd_{1-x}Bi_xMnO_3$ samples.

Doping Content (*x*)	**0.0**	**0.25**	**0.50**
Space group	Pbnm	Pbnm	Pbnm
Symmetry	Orthorhombic	Orthorhombic	Orthorhombic
***a* (Å)**	5.4240	5.4292	5.4321
***b* (Å)**	5.5570	5.6534	5.7712
***c* (Å)**	7.6648	7.6323	7.8536
$\alpha = \beta = \gamma$	90	90	90
Volume (Å^3)	231.03	234.27	236.25

Table 2. Bond length and bond angle between Mn–O–Mn and goodness of fit for $Nd_{1-x}Bi_xMnO_3$ samples.

Doping Content (*x*)	**0**	**0.25**	**0.50**
Mn–O_1 (Å)	1.9903 (5)	2.0407 (7)	1.9874 (4)
Mn–O_2 (Å)	2.0256 (5)	1.9685 (6)	1.9289 (3)
<Mn–O> (Å)	2.0080 (5)	2.0046 (6)	1.9581 (2)
Mn–O_1–Mn (°)	148.62(1)	138.45 (3)	148.58 (9)
Mn–O_2–Mn (°)	154.61(1)	149.36 (12)	154.08 (4)
<Mn–O–Mn> (°)	151.62 (1)	143.91 (2)	151.33 (6)
χ^2	1.038	2.581	3.812
R_P (%)	8.27	16.76	14.37
R_{WP} (%)	10.98	21.10	19.87
$<r_A>$ (Å)	1.109	1.142	1.175
τ	0.515	0.522	0.529

Note that $<r_A>$ are calculated from ionic radii r(Nd) = 1.109 Å and r(Bi) = 1.24 Å.

Figure 3 illustrates the 3D representation of the compound of $Nd_{1-x}Bi_xMnO_3$, which shows an octahedral MnO_6, constructed using Visualisation for Electronic and Structural Analysis (VESTA) software. In this structure, the compounds Nd^{3+}, Mn^{3+} cations and O^{2-} anions occupy the corner, body center and face center positions, respectively. The A site in the ABO_{3-} type perovskite structure occupied by

Nd/Bi cations was surrounded by 12 oxygen ions, while the octahedral MnO_6 formed by the position of Mn ions at the B site was surrounded by six oxygen ions.

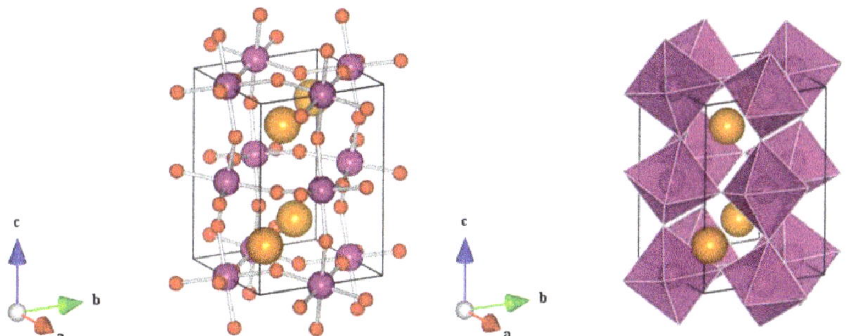

Figure 3. Crystallographic structure of $NdMnO_3$. Purple, orange and red-colored balls represent the Mn, Nd and O, respectively.

3.2. FTIR Spectra

Figure 4 shows the spectra for the compound of $NdMnO_3$ doped with Bi in the frequency range from 450 cm^{-1}–2000 cm^{-1}. From this figure, we can see that the most significant absorption bands are located at the range of 600 cm^{-1}–550 cm^{-1}. The absorption bands at this range correspond to the Mn–O stretching vibration, linked to the internal movement of a length change of the bounds Mn–O–Mn, associated with the octahedron MnO_6 attributed to a vibration characteristics of the perovskite-type ABO_3 [3,19]. The absorption peak band at the range of 600 cm^{-1}–550 cm^{-1} seem to shift towards lower wave number values for a structure having lower symmetry with decreasing Bi^{3+} ion content, according to the increase of e_g electrons in the anti-bonding orbital. Hence, the bond order would decrease with an increase in the number of e_g electrons. The lower frequency band can be related to a higher deformation mode of the MnO_6 octahedron [20,21]. The FTIR spectra confirm the formation of the $NdBiMnO_3$ sample.

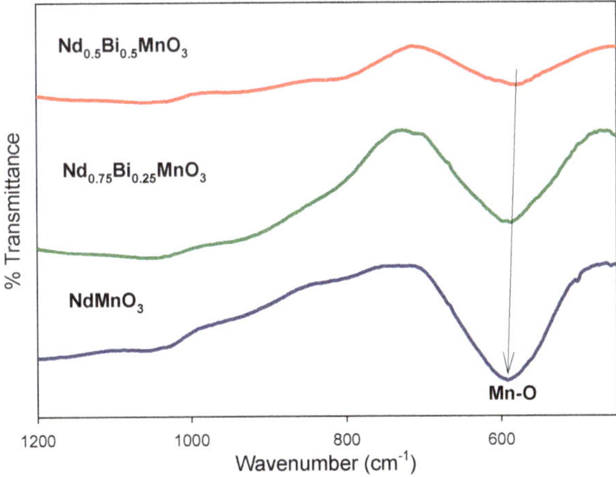

Figure 4. Fourier transform infrared spectroscopy (FTIR) spectra of $Nd_{1-x}Bi_xMnO_3$ (x = 0, 0.25 and 0.50).

3.3. Morphology

To understand the morphology of the samples, we performed FESEM with EDX and obtained the results, as shown in Figure 5. The images in Figure 5 reveal that the surface of the sample is non-homogeneous and the grains are agglomerated. The particles are also observed as non-spherical in shape. According to the image, the grain size of all samples are in the range of ~3.7 µm to ~5.6 µm, respectively. The grain sizes observed by FESEM are larger than those calculated by the Debye–Scherer formula. This can be explained by the fact that each particle observed by FESEM is formed by several crystallized grains. The results of EDX plot matches with a standard peak position of Nd, Bi, Mn and O from the previous study [19]. The EDX spectrum confirms the homogeneity of the samples.

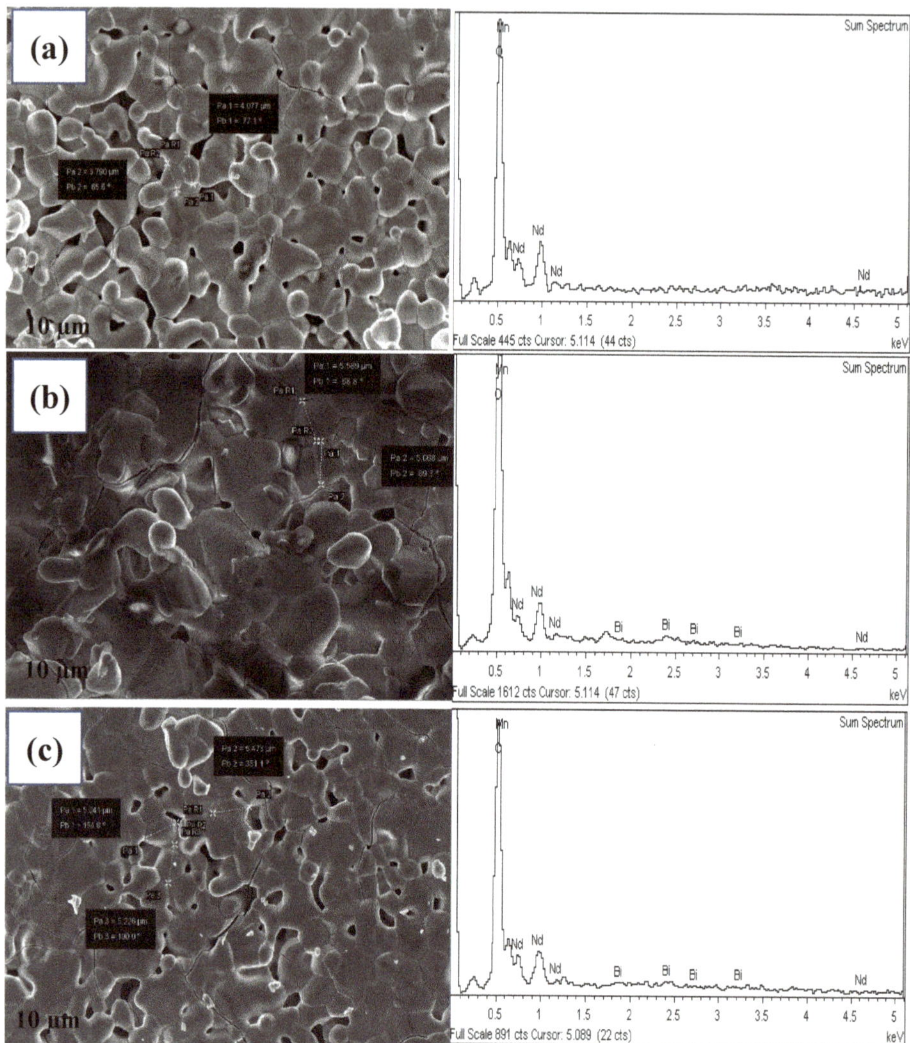

Figure 5. Field emission scanning electron microscopy (FESEM) micrograph and EDX spectra of $Nd_{1-x}Bi_xMnO_3$ (**a**) x = 0 (**b**) x = 0.25 and (**c**) x = 0.5.

3.4. AC Susceptibility Measurement

Figure 6 shows the temperature dependence of the AC susceptibility of the $Nd_{1-x}Bi_xMnO_3$ system on the real part χ' with the inset $d\chi'/dT$ versus the temperature range of 30–300 K. For x = 0 and x = 0.25, a definite and sharp peak was observed at 70 K and 72 K, respectively, associated to the ordering temperature of Mn spins [22]. The observed cusp depicts the PM to AFM transition, which is consistent with the results from previous studies [5,8,23]. For x = 0.50, two sharp peaks were observed, which were at 42 K and 57 K, respectively. The existence of two peaks indicates the magnetic inhomogeneity that may be induced by Bi substitution [12]. The T_C was determined using the minimum, while T_N was determined using the maximum point of differentiation, as shown in the inset of the graph in Figure 6. In the figure, all the three compounds exhibit a strong AFM with a competing weak FM, where T_N and T_C decreased with Bi substitution. The decreasing value of T_C shows the weakening of the FM behavior due to the Bi substitution. Temperature dependence of inverse susceptibility, $1/\chi'$, for all samples is plotted in Figure 7. The experimental data were fitted by using Curie–Weiss (C–W) equation, $\chi = C/(T - \theta)$, where C is Curie constant and θ is Curie–Weiss temperature. The red line is the best fit for C–W law in the paramagnetic region. The inverse susceptibility curve follows the C–W law at the higher temperature and starts to deviate from the linear fitting of C–W law at the lower temperature above T_C, which shows the existence of Griffiths phase (GP) in the samples [24,25]. The deviation shows that the C–W law is inapplicable in the temperature range between T_C and Griffiths temperature, T_G ($T_C < T < T_G$). This anomaly indicates the possible existence of ferromagnetic cluster within the paramagnetic region in the system which would enhance the total magnetic susceptibility of the samples [24–27]. The temperature ranges of Griffiths phase normalized with T_C were evaluated using the equation GP = $(T_G - T_C)/T_C$ and were found to increase as the substitution of Bi increased. The increasing values were due to an increase of magnetic inhomogeneity near T_C. Therefore, the substitution of Bi in $Nd_{1-x}Bi_xMnO_3$ system could enhance the appearance of Griffiths phase [24]. The values of T_C, T_N, T_G and GP are summarized in Table 3.

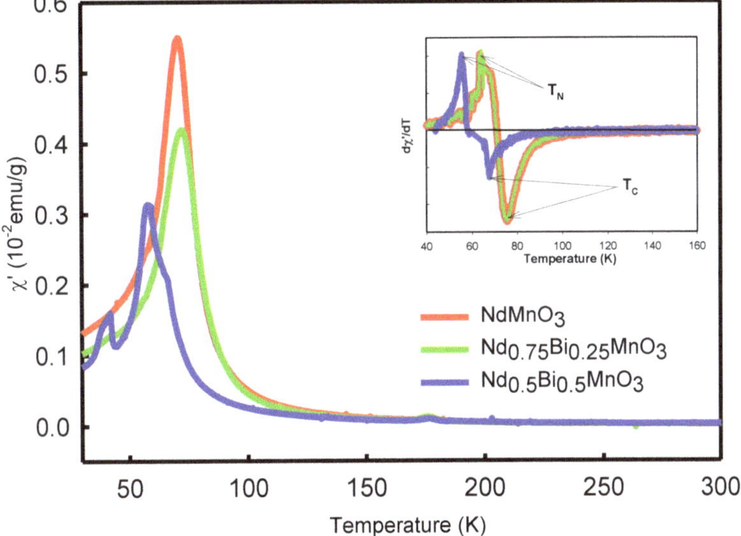

Figure 6. AC magnetic susceptibility against temperature curves for $Nd_{1-x}Bi_xMnO_3$.

Table 3. Parameters obtained from AC magnetic susceptibility studies of $Nd_{1-x}Bi_xMnO_3$ samples.

x	0	0.25	0.50
T_N (K)	64	64	55
T_C (K)	76	75	67
T_G (K)	181	198	182
GP (K)	138	164	171

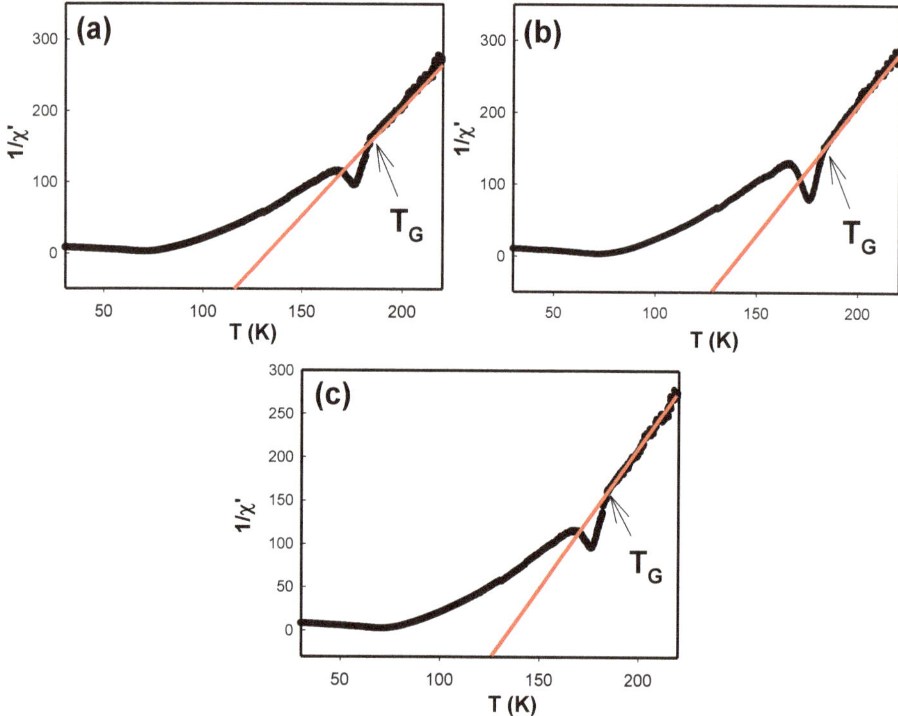

Figure 7. Temperature dependence of inverse susceptibility spectra of $Nd_{1-x}Bi_xMnO_3$ (**a**) x = 0, (**b**) x = 0.25 and (**c**) x = 0.5.

4. Conclusions

In summary, this study investigates the effect of Bi^{3+} substitution on structural, optical, morphological and magnetic properties of $Nd_{1-x}Bi_xMnO_3$ samples. The XRD analysis and Rietveld refinement show the pattern corresponding to the perovskite-type $NdMnO_3$, which crystallizes in the orthorhombic system with main peak at (121) hkl plane with the Pbnm space group. The structural parameter and cell volume of the samples increase as the Bi^{3+} content increases. In addition, the average ionic radii of A site, $<r_A>$, increased, hence τ also increased. The FTIR spectra show that active vibrational bands present at 600 cm^{-1}–550 cm^{-1}, which correspond to the metal oxide Mn–O stretching vibration of perovkite $NdMnO_3$. FESEM images show that the sample is non-spherical and somewhat agglomerated. The samples show a strong AFM behavior at low temperature with decreasing T_C due to the Bi substitution. Deviation of temperature dependence of inverse susceptibility curves shows the existence of Griffiths phase in these materials.

Author Contributions: Z.M. conceived and designed the experiment; I.S.I. performed the experiments supervised by Z.M. The manuscript was written by N.A.A. and Z.M. with contributions from N.B.M. and A.H. All authors have read and agreed to the published version of the manuscript.

Funding: This research was financially supported by the Ministry of Education Malaysia (MOE) and Universiti Teknologi MARA, grant number 600-IRMI/FRGS-RACER 5/3 (042/2019).

Acknowledgments: The authors would like to thanks the Faculty of Applied Sciences, Universiti Teknologi MARA for FESEM and FTIR characterization, and Center for Superconducting & Magnetic Novel Oxides of Universiti Teknologi MARA for ACS characterization.

Conflicts of Interest: The authors declare no conflict of interest.

References

1. Massa, N.E.; Campo, L.; Meneses, D.D.S.; Martínez-lope, M.J.; Alonso, J.A. High temperature far infrared dynamics of orthorhombic $NdMnO_3$; emissivity and reflectivity. *J. Phys. Condens. Matter.* **2013**, *25*, 235603. [CrossRef] [PubMed]
2. Renwen, Q.; Zhe, L.; Jun, F. Influence of Bi^{3+} doping on electronic transport properties of $La_{0.5x}Bi_xCa_{0.5}MnO_3$ manganites. *Phys. B Phys. Condens. Matter.* **2011**, *406*, 1312–1316.
3. Hernández, E.; Sagredo, V.; Delgado, G.E. Synthesis and magnetic characterization of $LaMnO_3$ nanoparticles. *Rev. Mex. Fis.* **2015**, *61*, 166–169.
4. Ahmadvand, H.; Salamati, H.; Kameli, P. The effect of grain boundaries on the domain wall dynamics in $Pr_{1-x}Ag_xMnO_3$ manganites. *J. Appl. Phys.* **2010**, *107*, 083913. [CrossRef]
5. Nandy, A.; Roychowdhury, A.; Das, D.; Pradhan, S.K. Structural and magnetic characterizations of undoped and K-doped $NdMnO_3$ single crystals synthesized by sol-gel route: A comparative study. *Powder Technol.* **2014**, *254*, 538–547. [CrossRef]
6. Chatterji, T.; Ouladdiaf, B.; Bhattacharya, D. Neutron diffraction investigation of the magnetic structure and magnetoelastic effects in $NdMnO_3$. *J. Phys. Condens. Matter.* **2009**, *21*, 1–6. [CrossRef]
7. Liu, K.; Wu, X.W.; Ahn, K.H.; Sulchek, T.; Chien, C.L.; Xiao, J.Q. Charge ordering and magnetoresistance in $Nd_{1-x}Ca_xMnO_3$ due to reduced double exchange. *Phys. Rev. B* **1996**, *54*, 3007. [CrossRef]
8. Wu, S.Y.; Hwang, S.R.; Li, W.; Lee, K.C.; Lynn, J.W.; Liu, R.S. Jahn-Teller distortion, charge ordering, and magnetic transitions in Ca-Doped $NdMnO_3$. *J. Phys.* **2000**, *38*, 354–359.
9. Arifin, M.; Ibrahim, N.; Mohamed, Z.; Yahya, A.K.; Khan, N.A.; Khan, M.N. Revival of metal-insulator and ferromagnetic-paramagnetic transitions by Ni substitution at Mn site in charge-ordered monovalent doped $Nd_{0.75}Na_{0.25}MnO_3$ manganites. *J. Supercond. Nov. Magn.* **2018**, *31*, 2851–2868. [CrossRef]
10. Tang, T.; Tien, C.; Hou, B.Y. Electrical transport and magnetic properties of $Nd_{1-x}Na_xMnO_3$ manganites. *J. Alloys Compd.* **2008**, *461*, 42–47. [CrossRef]
11. Chiba, H.; Atou, T.; Syono, Y. Magnetic and Electrical Properties of $Bi_{1-x}Sr_xMnO_3$: Hole-Doping Effect on Ferromagnetic Perovskite $BiMnO_3$. *J. Solid State Chem.* **1997**, *132*, 139–143. [CrossRef]
12. Ghani, M.A.; Mohamed, Z.; Yahya, A.K. Effects of Bi substitution on magnetic and transport properties of $La_{0.7-x}Bi_xAg_{0.3}MnO_3$ ceramics. *J. Supercond. Nov. Magn.* **2012**, *25*, 2395–2402. [CrossRef]
13. Zhao, Y.D.; Park, J.; Jung, R.J.; Noh, H.J.; Oh, S.J. Structure, magnetic and transport properties of $La_{1-x}Bi_xMnO_3$. *J. Magn. Magn. Matter.* **2004**, *280*, 404–411. [CrossRef]
14. Sugawara, F.; Iiida, S. Magnetic properties and crystal distortions of $BiMnO_3$ and $BiCrO_3$. *J. Phys. Soc. Jpn.* **1968**, *25*, 1553–1558. [CrossRef]
15. Lalitha, G.; Reddy, P.V. Low temperature resistivity anomalies in bismuth doped manganites. *J. Alloys Compd.* **2010**, *494*, 476–482. [CrossRef]
16. Alonso, J.A.; Martínez-Lope, M.J.; Casais, M.T.; Fernández-Díaz, M.T. Evolution of the Jahn-Teller distortion of MnO_6 octahedra in $RMnO_3$ perovskites (R = Pr, Nd, Dy, Tb, Ho, Er, Y): A neutron diffraction study. *Inorg. Chem.* **2000**, *39*, 917–923. [CrossRef]
17. Srivastava, S.R.S.K. Magnetic properties of $Nd_{1-x}Ag_xMnO_3$ compounds. *J. Phys. Condens. Matter.* **2008**, *20*, 505212. [CrossRef]
18. Sangale, M.D.; Gaikwad, D.N.; Sonawane, D.V.; Suryavanshi, D.M. Synthesis and characteristic properties of perovskite-type $NdMnO_3$ nanocrystal materials via a co-precipitation method. *J. Chem. Phys. Sci.* **2018**, *7*, 291–296.
19. Somvanshi, A.; Husain, S. Study of structural, dielectric and optical properties of $NdMnO_3$. *AIP Conf. Proc.* **2018**, *1953*, 1–5.
20. Turky, A.O.; Rashad, M.M.; Hassan, A.M.; Elnaggar, E.M.; Bechelany, M. Optical, electrical and magnetic properties of lanthanum strontium manganite $La_{1-x}Sr_xMnO_3$ synthesized through the citrate combustion method. *Phys. Chem. Chem. Phys.* **2017**, *19*, 6878–6886. [CrossRef]
21. Mohamed, Z.; Shahron, I.S.; Ibrahim, N.; Maulud, M.F. Influence of ruthenium doping on the crystal structure and magnetic properties of $Pr_{0.67}Ba_{0.33}Mn_{1-x}Ru_xO_3$ Manganites. *Crystal* **2020**, *10*, 295. [CrossRef]
22. Wu, S.Y.; Kuo, C.M.; Wang, H.Y.; Li, W.-H.; Lee, K.C. Magnetic structure and spin reorientation of the Mn ions in $NdMnO_3$. *J. Appl. Phys.* **2000**, *87*, 5822–5824. [CrossRef]

23. Asmira, N.; Ibrahim, N.; Mohamed, Z.; Yahya, A.K. Effect of Cr^{3+} substitution at Mn-site on electrical and magnetic properties of charge ordered $Bi_{0.3}Pr_{0.3}Ca_{0.4}MnO_3$ manganites. *Phys. B Condens. Matter.* **2018**, *544*, 34–46. [CrossRef]
24. Manjunatha, S.O.; Rao, A.; Poornesh, P.; Lin, W.J.; Kuo, Y.K. Magnetic inhomogeneity and Griffiths phase in Bi substituted $La_{0.65-x}Bi_xCa_{0.35}MnO_3$ manganites. *Phys. B Condens. Matter.* **2016**, *498*, 82–91. [CrossRef]
25. Zheng, X.; Gao, T.; Jing, W.; Wang, X.Y.; Liu, Y.S.; Chen, B.; Dong, H.L.; Chen, Z.Q.; Cao, S.X.; Cai, C.B.; et al. Evolution of Griffiths phase and spin reorientation in perovskite manganites. *J. Magn. Magn. Mater.* **2019**, *491*, 165611. [CrossRef]
26. Pękała, M.; Szydłowska, J.; Pękała, K.; Drozd, V. Griffiths like phase in nanocrystalline manganite $La_{0.50}Ca_{0.50}MnO_3$ studied by magnetic susceptibility and electron spin resonance. *J. Alloys Compd.* **2016**, *685*, 237–241. [CrossRef]
27. Elyana, E.; Mohamed, Z.; Kamil, S.A.; Supardan, S.N.; Chen, S.K.; Yahya, A.K. Revival of ferromagnetic behavior in charge-ordered $Pr_{0.75}Na_{0.25}MnO_3$ manganite by ruthenium doping at Mn site and its MR effect. *J. Solid State Chem.* **2018**, *258*, 191–200. [CrossRef]

Article

Simulating the Hysteretic Characteristics of Hard Magnetic Materials Based on $Nd_2Fe_{14}B$ and $Ce_2Fe_{14}B$ Intermetallics

Natalia B. Kolchugina [1,*], Mark V. Zheleznyi [1,2], Aleksandr G. Savchenko [2], Vladimir P. Menushenkov [2], Gennadii S. Burkhanov [1], Yurii S. Koshkid'ko [1,3], Jacek Ćwik [3], Nikolai A. Dormidontov [1], Katerina Skotnicova [4], Miroslav Kursa [4] and Pavel A. Prokofev [1,5]

1 Baikov Institute of Metallurgy and Materials Science, Russian Academy of Sciences, Leninskii pr. 49, 119334 Moscow, Russia; markiron@mail.ru (M.V.Z.); gburkhanov@imet.ac.ru (G.S.B.); y.koshkidko@intibs.pl (Y.S.K.); ontip@mail.ru (N.A.D.); pav3387@yandex.ru (P.A.P.)
2 National University of Science and Technology MISiS, Leninskii pr. 4, 119991 Moscow, Russia; algsav@gmail.com (A.G.S.); menushenkov@gmail.com (V.P.M.)
3 Institute of Low Temperature and Structure Research, Polish Academy of Sciences, ul. Okólna 2, 50-422 Wroclaw, Poland; j.cwik@intibs.pl
4 VSB—Technical University of Ostrava, Faculty of Materials Science and Technology, 17. listopadu 2172/15, 70800 Ostrava, Czech Republic; Katerina.Skotnicova@vsb.cz (K.S.); Miroslav.Kursa@vsb.cz (M.K.)
5 JSC Spetsmagnit, Dmitrovskoe sh. 58, 127238 Moscow, Russia
* Correspondence: nkolchugina@imet.ac; Tel.: +7-(499)135-96-15

Received: 12 May 2020; Accepted: 14 June 2020; Published: 17 June 2020

Abstract: The $Ce_2Fe_{14}B$ intermetallic, like $Nd_2Fe_{14}B$, has the tetragonal $Nd_2Fe_{14}B$-type structure (space group $P4_2/mnm$), in which Ce ions have a mixed-valence state characterized by the coexistence of trivalent $4f^1$ and tetravalent $4f^0$ electron states. Despite the fact that the saturation magnetization, magnetic anisotropy field, and Curie temperature of the $Ce_2Fe_{14}B$ intermetallic are substantially lower than those of $Nd_2Fe_{14}B$ and $Pr_2Fe_{14}B$, $Ce_2Fe_{14}B$ retains the capacity of being able to be used in the manufacturing of rare-earth permanent magnets. Moreover, at low temperatures, the anisotropy field of $Ce_2Fe_{14}B$ is higher than that of $Nd_2Fe_{14}B$, and $Ce_2Fe_{14}B$ does not undergo the spin-reorientation transition. In this respect, studies of (Nd, Ce)-Fe-B alloys, which are intended for the improvement of the service characteristics-to-cost ratio, are very relevant. A model and algorithm for calculating the hysteresis loops of uniaxial hard magnetic materials with allowance for the K_1 and K_2 ($K_2 > 0$ and $K_1 > 0$ and $K_1 < 0$) magnetic anisotropy constants were developed and allowed us to obtain data on their effect on the parameters of hysteresis loops for a wide temperature range (0–300 K). The simulation and analysis of hysteresis loops of the quasi-ternary intermetallics $(Nd_{1-x}Ce_x)_2Fe_{14}B$ (x= 0–1) was performed. Results of the simulation indicate that the alloying of the $Nd_2Fe_{14}B$ intermetallic with Ce to $x = 0.94$ (1) does not completely eliminate the negative effect of spin-reorientation phase transition on the residual magnetization of the $(Nd_{1-x}Ce_x)_2Fe_{14}B$ intermetallic and (2) slightly decreases the slope of magnetization reversal curve.

Keywords: R-Fe-B intermetallics; cerium; permanent magnets; simulation; magnetic anisotropy constant; hysteresis loop; coercive force; residual magnetization

---

1. Introduction

In recent years, the low cost of Ce-containing Nd-Fe-B permanent magnets, in some areas being a potential alternative to those based on expensive rare-earth elements (Nd, Pr, Dy, Tb), has stimulated considerable research efforts [1,2].

Similar to the other R-Fe-B systems, the Ce-Fe-B system is characterized by the formation of the ternary intermetallic compound having a 2:14:1 stoichiometry and tetragonal $Nd_2Fe_{14}B$-type structure (space group $P4_2/mnm$) [3]. The unit cell contains four formula units comprising 68 atoms: there are six crystallographic iron sites ($16k_1$, $16k_2$, $8j_1$, $8j_2$, 4c, 4e), two rare-earth metal sites (4*f*, 4*g*), and one boron site (4*g*). Table 1 shows the lattice parameters of the $R_2Fe_{14}B$ compounds with R = Nd, Pr, and Ce and their principal magnetic characteristics. This shows that both the *a* and *c* lattice parameters of the $Ce_2Fe_{14}B$ compound are slightly lower than those of the $Nd_2Fe_{14}B$ and $Pr_2Fe_{14}B$ compounds [4].

Table 1. Saturation magnetization I_s, magnetic anisotropy field H_A, lattice parameters *a* and *c*, and Curie temperature T_C of the $R_2Fe_{14}B$ compounds with R = Ce, Pr, Nd at room temperature.

Compound	I_s, T	H_A, kOe	T_C, K	*a*, nm	c, nm	Reference
$Nd_2Fe_{14}B$	1.61	73	585	0.880	1.220	[4,5]
$Pr_2Fe_{14}B$	1.56	87	565	0.880	1.223	[4,6]
$Ce_2Fe_{14}B$	1.17	26	424	0.876	1.211	[4,7]

The anisotropy of the $Nd_2Fe_{14}B$ compound is dominated by the rare-earth atoms that occupy two inequivalent sites, 4*f* and 4*g*, of the tetragonal structure [8,9]. One Nd site (*g*) strongly prefers the [001] direction at ambient temperature and dictates the macroscopic easy-axis direction. The other Nd site (*f*) (containing half of all the Nd atoms) reduces the intrinsic stability by favoring alignment along [110]-type directions (basal plane). The results indicate that coercivity may be enhanced by preferential chemical doping of Nd *f* sites. $Nd_2Fe_{14}B$ is characterized by the uniaxial state at temperatures above the spin-reorientation temperature T_{sr}. $Ce_2Fe_{14}B$ exhibits the uniaxial magnetic anisotropy over the whole temperature range of ferromagnetic ordering [10]. Its uniaxial anisotropy is higher than that of $R_2Fe_{14}B$ with R =La, Lu, Y [10], but lower than that of $Nd_2Fe_{14}B$ at temperatures substantially higher than the spin-reorientation temperature. It was predicted [11] that, theoretically, Ce atoms in the $(Nd_{1-x}Ce_x)_2Fe_{14}B$ compounds occupy the 4g positions (large in volume); this is explained by atomic size effects. However according to [8], 4f position (smaller in volume) is preferred for the Ce atoms.

Taking into account the fact that Ce prefers 4f positions, the progressive substitution of Ce for Nd should decrease the easy magnetization axis (EMA) cone opening. Thus, the formation of a uniaxial state can occur even in the presence of a small amount of Nd in $(Nd_{1-x}Ce_x)_2Fe_{14}B$. At present, no experimental data confirming the assumption are available. In turn, when assuming that Ce atoms prefer 4g positions, it is possible to conclude that the cone opening will be smaller for low substitutions of Ce for Nd. Because of this, T_{sr} varies slightly for low Ce contents. A small change in T_{sr} for the Ce substitution to x = 0.3 was observed in [12]. According to the data from [8], the magnetocrystalline anisotropy of $Nd_2Fe_{14}B$ remains when Ce substitutes for Nd up to 20%. The further increase in the Ce content decreases the uniaxial magnetic anisotropy energy.

The uniaxial anisotropy found in $Ce_2Fe_{14}B$ is attributed mainly to the magnetism of Fe [8]. Alloying Ce on R-sites in $(Nd_{1-x}Ce_x)_2Fe_{14}B$ barely affects the Fe moments. Nd magnetization changes orientation at x > 0.5, with a larger antiferromagnetic moment on 4f sites compared to 4g sites. Such a transition of the Nd magnetic state causes an overall reduction of the net magnetization of the cell, and is the reason for instability at higher percentage Ce [11]. The magnetic moment of rare-earth sublattice is mainly determined by Nd atoms. The temperature behavior of magnetization of $R_2Fe_{14}B$ with R = Nd, Ce was studied in [12].

The metallurgical behavior, heat treatment conditions, fundamental characteristics, and magnetic properties of the $Nd_2Fe_{14}B$ and $Ce_2Fe_{14}B$ intermetallics differ substantially although the compounds have the same crystal structure. The existence of the $CeFe_2$ phase determines the principal difference in the ternary Nd-Fe-B and Ce-Fe-B phase diagrams [3,13]. Cerium ions in the $Ce_2Fe_{14}B$ intermetallic have a mixed-valence state, namely, the trivalent $4f^1$ and tetravalent $4f^0$ electron states coexist [14].

According to data from [15], the hysteretic properties of Nd-Ce-Fe-B magnets decrease as the Ce content increases. However, the squareness of the magnetization reversal curve remains high [16–18], and the magnetic characteristics remain adequate when the Ce content does not exceed 10%.

Usually, it is assumed that the decrease in the magnetic characteristics of Ce-containing $Nd_2Fe_{14}B$ alloys is due to the lower magnetic properties of $Ce_2Fe_{14}B$ as compared to those of $Nd_2Fe_{14}B$ (see Table 1). However, according to data from [18,19], an anomalous increase in the coercive force was found by studying the effect of Ce substitution for Nd on the magnetic properties and microstructure of sintered magnets. Pathak et al. [20] reported that the substitution of 20% Ce for Nd in the ternary $Nd_2Fe_{14}B$ alloy allowed the authors to reach a sufficiently high coercive force (H_{ci} = 10 kOe), which exceeds that of $Nd_2Fe_{14}B$ (H_{ci} = 8.3 kOe).

Currently, the development of high-coercivity, high-performance permanent magnets operating in a wide temperature range, in particular at low and cryogenic temperatures, is of importance. Nd-Fe-B magnets are unsuitable for operation at such temperatures. The EMA of the $Nd_2Fe_{14}B$ compound at 4.2 K is in the [110] plane and makes the angle $\theta \approx 30°$ with the c axis. As the temperature increases, the transition to the collinear structure takes place at the spin-reorientation temperature T_{sr} = 135–138 K [21]. Below this temperature, the magnetic moment deviates from the c axis, the first magnetic anisotropy constant K_1 passes zero and changes the sign from positive to negative, whereas the second magnetic anisotropy constant remains positive ($K_2 > 0$). As a result, below T_{sr}, the experimental magnetization reversal curves in negative magnetic fields exhibit a bending, which increases with decreasing temperature. In this case, the residual magnetization and maximum energy product decrease abruptly.

At room temperature, the magnetic properties of $Ce_2Fe_{14}B$ are substantially lower than those of $Nd_2Fe_{14}B$, whereas, at cryogenic temperatures, the magnetic anisotropy field of $Ce_2Fe_{14}B$ is markedly higher than that of $Nd_2Fe_{14}B$. Moreover, it is of importance that $Ce_2Fe_{14}B$ does not have a spin-reorientation transition. Thus, it is reasonable to expect that the partial substitution of Ce for Nd in the $Nd_2Fe_{14}B$ compound can lead to the improvement of the hysteretic characteristics of permanent magnets based on the quasi-ternary $(Nd, Ce)_2Fe_{14}B$ intermetallics.

Data on the effect of Ce substitution for Nd on the spin-reorientation transition temperature of the $Nd_{2-x}Ce_xFe_{14}B$ single crystals are available in [22], where the evolution of the spin-reorientation temperature as a function of the Ce concentration up to x = 0.4 is considered. The spin-reorientation temperature decreases by only about 6% when Ce substitutes for 38% of Nd. It is shown that T_{sr} is suppressed much more rapidly for higher x. It is likely that the population of REM sites (4g or 4f) is responsible for the spin-reorientation [22].

Taking into account the absence of systematic data on this problem in the literature, the present study is aimed at the simulation of hysteretic properties of the $(Nd_{1-x}Ce_x)_2Fe_{14}B$ intermetallics in order to determine the optimum alloyed compositions. The simulation and analysis of hysteresis loops of the $(Nd_{1-x}Ce_x)_2Fe_{14}B$ (x = 0–1) compounds are performed for a temperature range of 0 to 300 K.

2. Algorithm and Model for Calculating the Hysteresis Loops of Hard Magnetic Materials with the Uniaxial Tetragonal Lattice

The resulting magnetization vector I_s at each point of a crystalline ferromagnet is simultaneously oriented along certain crystallographic directions (EMAs). In order to rotate I_s to another direction, magnetic field $\boldsymbol{H}$ should be applied along this direction, and work should be done. This work makes sense of the anisotropy energy E_a, which, for magnets with the uniaxial tetragonal lattice, is given by the expression:

$$E_a = -K_1 \, cos^2(\varphi) - K_2 \, cos^4(\varphi), \tag{1}$$

where K_1 and K_2 are the first and second magnetic anisotropy constants, respectively, and φ_a is the angle made by the EMA and I_s. The anisotropy constants are proportional to the work, which should be done to rotate the magnetization from the EMA direction to the hard magnetization axis direction.

To take into account the cooperative effect of magnetic anisotropy and magnetic field, we considered a single crystal in the form of a plain disk oriented along a certain crystallographic plane. When the magnetic field is applied along the plane, the magnetization is also within the plane. It is assumed that the magnetization of the body is uniform and domains are absent.

When applying magnetic field H, the position of I_s also determined by the magnetic field energy:

$$E_H = -H\, I_s\, cos(\varphi), \tag{2}$$

where φ_H is the angle made by the vectors. In the end, I_s takes a position corresponding to the minimum summary energy:

$$E = E_a + E_H \tag{3}$$

In terms of the model and algorithm [23] (program Hysteresis developed by Associate Prof. V.L. Stolyarov) used for the simulation of magnetic hysteresis loops of $(Nd_{1-x}Ce_x)_2Fe_{14}B$ intermetallics with the uniaxial tetragonal lattice, the following initial parameters are inputted: α is the angle made by an arbitrary plane and the X-axis [100] and the external magnetic field $\boldsymbol{H}$ is applied in this plane, θ_H is the angle made by the H field direction and Z-axis [001]. The following parameters are counted and outputted: θ is the angle made by the Z-axis [0001] and EMA, φ_a is the angle made by the EMA and $\boldsymbol{I}_s$; and angle $\varphi_{H.}$

Calculation of the position of the magnetization vector $\boldsymbol{I}_s$ for an arbitrary vector $\boldsymbol{H}$, at which the total energy is minimal (Equation (3)), allows us to calculate the projection of the magnetization on the field direction $I = I_s\cos(\varphi_H)$ and to construct the magnetization curve $I = I(\mathrm{H})$.

3. Results and Discussion

3.1. Determination of Magnetic Anisotropy Constants, Normalized Ratio of Anisotropy Constants ($K_2/|K_1|$), and Angle of the EMA cone in Calculating Hysteresis Loops of the $(Nd_{1-x}Ce_x)_2Fe_{14}B$ Intermetallics with $0 \le x \le 1$

Temperature dependences of the magnetic anisotropy constants K_1 and K_2 of $Nd_2Fe_{14}B$ in a temperature range of 0–500 K are available in [24]. The sign of K_1 alternates at the spin-reorientation transition temperature T_{sr} = 135 K. Below this temperature, the preferred direction of EMA begins to deviate from the c axis direction (Z-axis [001]) of the tetragonal crystal lattice, and the angle (θ) of the EMA cone for each temperature is given by the expression:

$$sin^2(\theta(T)) = -\frac{K_1(T)}{2\,K_2(T)} \tag{4}$$

At 4.2 K, this angle reaches ~30° [24].

The temperature dependences of the magnetic anisotropy constant K_1 and K_2 of $Ce_2Fe_{14}B$ in a temperature range of 0–300 K are available in [25]. Unlike the magnetic anisotropy constants of $Nd_2Fe_{14}B$, K_1 and K_2 of $Ce_2Fe_{14}B$ remain positive within the 0–300 Ktemperature range. The absolute values of K_1 and K_2 of $Ce_2Fe_{14}B$ are substantially lower than those of $Nd_2Fe_{14}B$.

To calculate the anisotropy constants of the quasi-ternary intermetallics $(Nd_{1-x}Ce_x)_2Fe_{14}B$ with $0 \le x \le 1$ at different temperatures, the literature data on the anisotropy constants of $Nd_2Fe_{14}B$ and $Ce_2Fe_{14}B$ and the following linear expressions were used:

$$K_1(T,\,x) = (1-x)\,K_1(T)_{Nd_2Fe_{14}B} + x\,K_1(T)_{Ce_2Fe_{14}B} \tag{5}$$

$$K_2(T,\,x) = (1-x)\,K_2(T)_{Nd_2Fe_{14}B} + x\,K_2(T)_{Ce_2Fe_{14}B} \tag{6}$$

The ratio of anisotropy constants K_2 to K_1 and the θ angle (made by the EMA and c axis) were calculated by the expressions:

$$\frac{K_2(T,\,x)}{\left|K_1(T,\,x)\right|} \tag{7}$$

at $K_1(T,\ x) < 0$

$$\theta(T) = arcsin\left(\sqrt{-\frac{K_1(T)}{2\,K_2(T)}}\right) \tag{8}$$

at $K_1(T,\ x) \geq 0$

$$\theta(T,\ x) = 0 \tag{9}$$

The following composition ranges of $(Nd_{1-x}Ce_x)_2Fe_{14}B$ were considered: (1) x = 0, 0.05, 0.10, 0.15, 0.20, 0.25, 0.30, 0.35, 0.40, 0.45, 0.50, 0.55, 0.60, 0.65, 0.70, 0.75; (2) x = 0.80, 0.85; 0.90, 0.91, 0.92 and (3) x = 0.93, 0.94, 0.95, 0.96, 0.97, 0.98, 0.99, 1.00.

The calculated temperature dependences of the magnetic anisotropy constants, their ratio, and temperature dependences of the θ angle for $(Nd_{1-x}\ Ce_x)_2Fe_{14}B$ with x = 0.05, 0.90, and 0.97 are given in Figure 1, as an example. The compositional dependences of the spin-reorientation temperature (T_{SR}) of $(Nd_{1-x}Ce_x)_2Fe_{14}B$ are given in Figure 2.

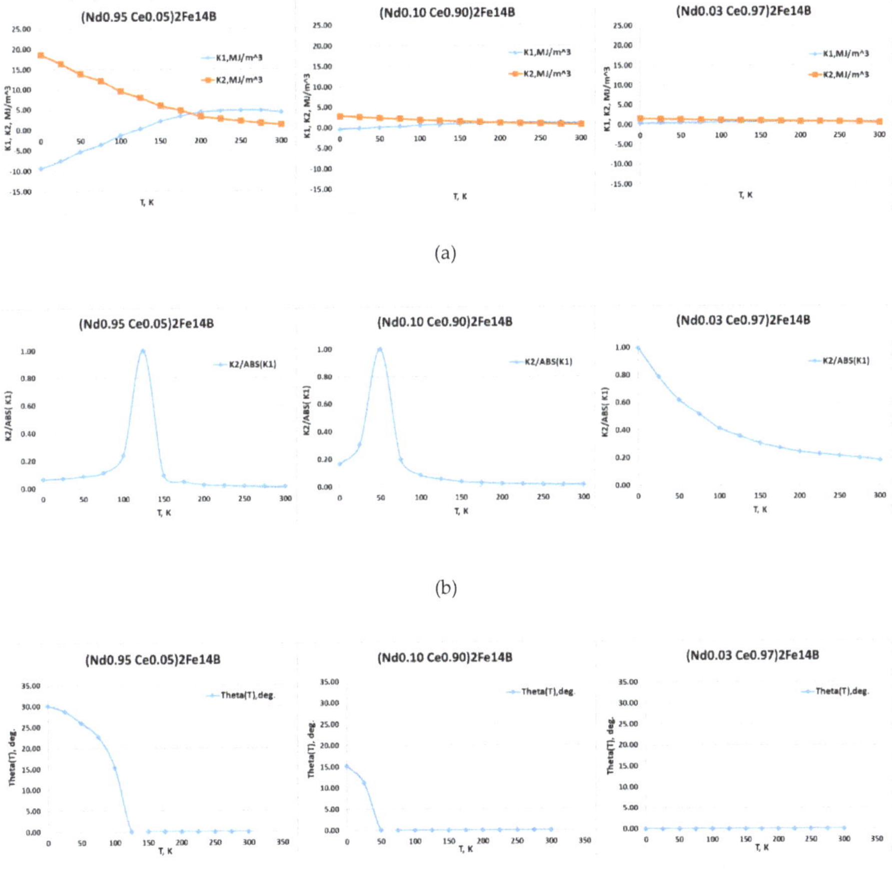

Figure 1. Temperature dependences of the (**a**) K_1 and K_2 magnetic anisotropy constants, (**b**) normalized $K_2/|K_1|$ ratio, and (**c**) θ angle for $(Nd_{1-x}\ Ce_x)_2Fe_{14}B$ with x = 0.05 (first composition range), 0.90 (second composition range), and 0.97 (third composition range).

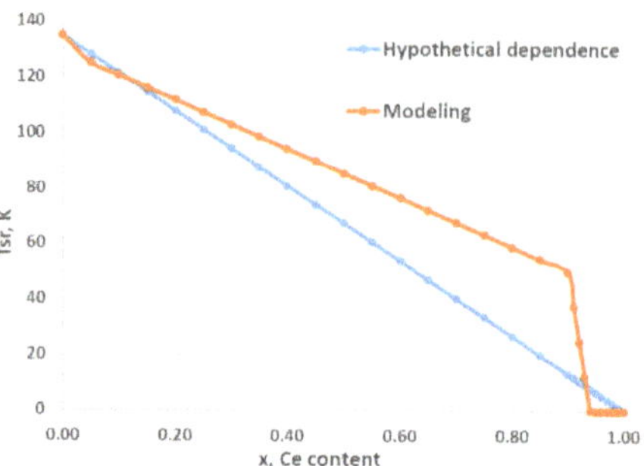

Figure 2. Compositional dependences of spin-reorientation temperature (T_{sr}) of $(Nd_{1-x}Ce_x)_2Fe_{14}B$: (blue) hypothetical (linear) trend and (red) results of simulation (this work).

It is seen from Figures 1 and 2 that, as Ce substitutes for Nd in the range x = 0–0.75 (first composition range), the values of K_1 and K_2 constants decrease on average by ~75% and ~99%, respectively. The temperature corresponding to the maximum of the normalized $K_2/|K_1|$ ratio decreases from 135 to 63 K. The temperature T_{sr} also decreases from 135 to 63 K. The highest value of the θ angle at 0 K decreases from 30.15° (x = 0) to 24.77° (x = 0.75).

As the Ce content in $(Nd_{1-x}Ce_x)_2Fe_{14}B$ increases in the range x = 0.80–0.92 (second composition range), the K_1 and K_2 values additionally decrease on average by ~61% and ~63%, respectively. In turn, the temperature corresponding to the maximum of normalized $K_2/|K_1|$ ratio shifts from 59 to 25 K. The temperature T_{sr} also decreases from 59 to 25 K. The largest value of the θ angle at 0 K monotonically decreases from 23.12° (x = 0.80) to 11.03° (x = 0.92).

It is seen from the dependences given in Figures 1 and 2 that, as the cerium content in the $(Nd_{1-x}Ce_x)_2Fe_{14}B$ intermetallic increases from x = 0.93 to x = 1.00, the anisotropy constant K_2 additionally decreases by an average ~35%, whereas the K_1 constant changes the sign from negative to positive. The T_{sr} temperature decreases from 13 to 0 K as the cerium content increases to x = 0.94. The temperature corresponding to the maximum of normalized $K_2/|K_1|$ ratio also decreases from 13 to 0 K. The highest value of the θ angle at 0 K decreases from 7.34° to 0°(at $x \geq 0.94$).

3.2. Simulation of Magnetization Curves and Hysteresis Loops of $(Nd_{1-x}Ce_x)_2Fe_{14}B$ with x = 0.00–1.00

The simulated hysteresis loops for $(Nd_{1-x}Ce_x)_2Fe_{14}B$ with x = 0–1.00 show that, as the applied magnetization reversing field reaches the coercive force H_c (at which the abrupt overturn of I_S takes place), the decrease in the hysteresis loop squareness is observed; the "rounding" becomes more substantial as the temperature decreases. Below T_{sr}, the decrease starts in the positive fields. Figure 3 shows the temperature dependences of the normalized and unnormalized residual magnetization (I_r/I_s and I_r) for the three composition ranges of the $(Nd_{1-x}Ce_x)_2Fe_{14}B$ compounds.

For all quasi-ternary compositions, the I_r/I_s ratio monotonically decreases with decreasing temperature. This is caused by the deviation of I_s from the EMA in the applied magnetic field H and the transition to the EMA cone below T_{sr}. Only for the high-cerium contents ($x \geq 0.94$), the I_r/I_s ratio remains unchanged and equal to 0.9999 (<1.0000) for a certain temperature range.

For the composition range x = 0–0.75 (Figure 3a), as the temperature decreases from 300 K to $T_{sr}(x)$, I_r/I_s remains unchanged and equal to 0.9999 for all these compositions. This means that, after saturation, the magnetization remains parallel to the EMA and Z-axis as the external field decreases to zero. The monotonic progressive decrease in I_r/I_s is observed simultaneously with decreasing temperature below $T_{sr}(x)$ and Ce content x. For each Ce content x, the temperature dependences $I_r(T, x)$ have the maximum value in the range of $T_{sr}(x)$ (Figure 3b). The monotonic shift of the I_r maximum to low temperatures from $I_r(T = 135\ K, x = 0)$ =191.6 A m^2/kg to $I_r(T = 102\ K, x = 0.75)$ = 161.5 A m^2/kg correlates with a similar shift of the maximum of normalized $K_2/|K_1|$ ratio and a decrease in $T_{sr}(x)$ with increasing cerium content x (Figures 1 and 2). The maximum in these dependences is related to the competition of two physical phenomena, such as the monotonic increase in the total magnetic moment of $R_2Fe_{14}B$ intermetallics (where R = Nd or Ce) with decreasing temperature and the spin-reorientation below the T_{sr} temperature.

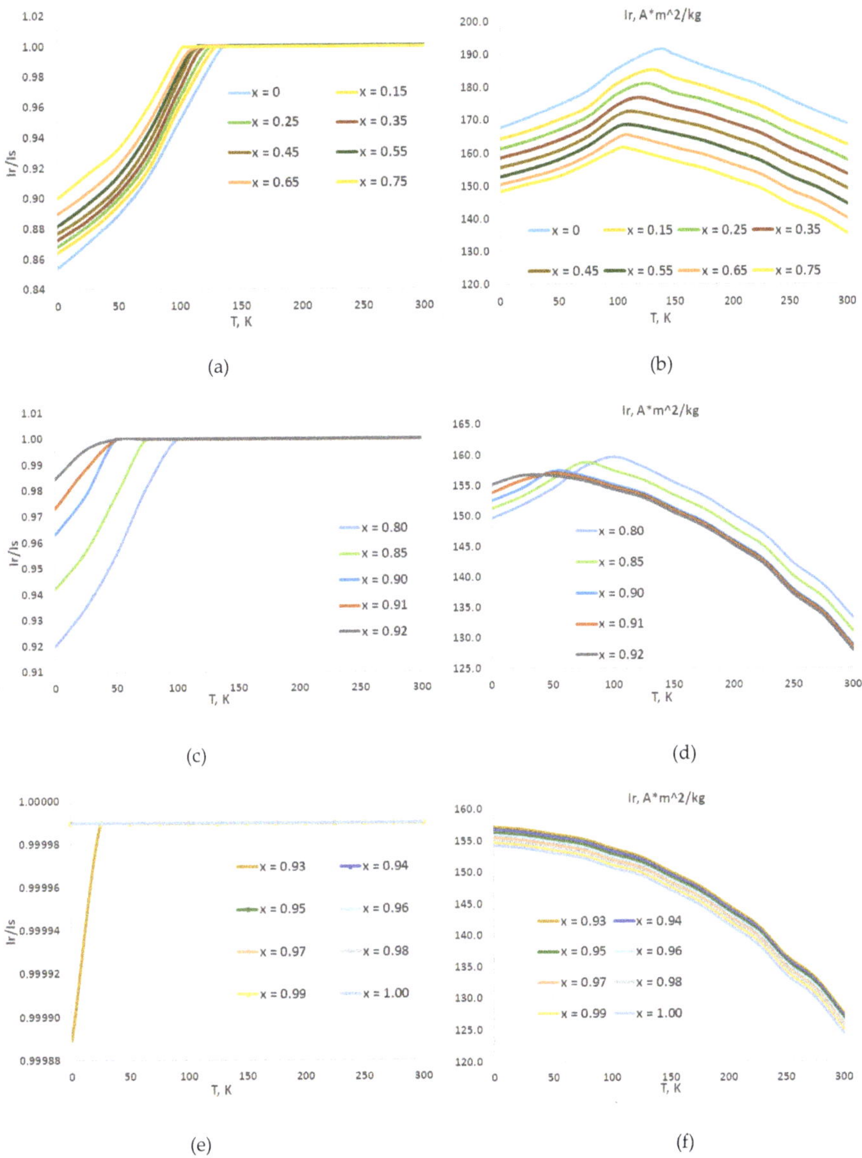

Figure 3. Temperature dependences of the normalized and unnormalized residual magnetization (I_r/I_s and I_r) of the $(Nd_{1-x}Ce_x)_2Fe_{14}B$ compounds with different x: (**a**) and (**b**) 0, 0.15, 0.25, 0.35, 0.45, 0.55, 0.65, 0.75; (**c**) and (**d**) 0.80, 0.85, 0.90, 0.91, 0.92; (**e**) and (**f**) 0.93, 0.94, 0.95, 0.96, 0.97, 0.98, 0.99, 1.00.

For the composition range x = 0.80–0.92 (Figure 3c), as the temperature decreases from 300 K to $T_{sr}(x)$, I_r/I_s remains equal to 0.9999 for all these compositions, i.e., the magnetization after saturation remains parallel to the EMA and Z-axis as the magnetizing field decreases to zero. The monotonic and progressive decrease in I_r/I_s is observed with simultaneously decreasing temperature below $T_{sr}(x)$ and Ce content x. For each Ce content x, the temperature dependences $I_r(T, x)$ also have the maximum in the range of $T_{sr}(x)$ (Figure 3d). The monotonic shift of the I_r maximum to the low temperature range from $I_r(T = 100\ K, x = 0.80)$ =159.6 A m^2/kg to $I_r(T = 25\ K, x = 0.92)$ = 156.6 A m^2/kg correlates with the similar shift of the normalized $K_2/|K_1|$ ratio and a decrease in $T_{sr}(x)$ with increasing cerium content x (Figures 1 and 2). However, in this range of cerium concentrations, the temperature dependences of I_r overlap; the overlapping was not observed in the range x = 0–0.75 (Figure 3d). This interesting and anomalous change in I_r is also related to the "stronger" competition of a monotonic increase in the total magnetic moment of $R_2Fe_{14}B$ intermetallics (where R = Nd or Ce) with decreasing temperature and the spin-reorientation below T_{sr}. Owing to the nature of the change in magnetic properties, the composition range x = 0.80–0.92 is intermediate between neodymium-based intermetallics and cerium-based intermetallics $R_2Fe_{14}B$.

Figure 3e shows the temperature dependences of the normalized residual magnetization for the third composition range of the $(Nd_{1-x}Ce_x)_2Fe_{14}B$ intermetallics with x = 0.93–1.00. As the temperature decreases from 300 K to 0 K, I_r/I_s remains equal to 0.9999 for all alloys with $x \geq 0.94$, i.e., after saturation, the magnetization remains parallel to the EMA and Z-axis as the magnetizing field decreases to zero. Only for the alloys with $0.93 \leq x < 0.94$, the monotonic and progressive decrease in I_r/I_s is observed as the simultaneous decrease in temperature below $T_{sr}(x)$ and decrease in the Ce concentration take place. As x increases from 0.94 to 1.00, the considered maximum of I_r shifts to zero and degenerates owing to the decreasing effect of the spin-reorientation. This effect also correlates with a similar shift of the normalized $K_2/|K_1|$ ratio and a decrease in $T_{sr}(x)$ with increasing cerium concentration x (Figures 1 and 2).

The considered features of the change of the residual magnetization clearly manifest themselves in the concentration dependence at T = 0 K. Figure 4 shows that the composition range x = 0.80–0.92 is intermediate between neodymium-based intermetallics and cerium-based intermetallic $R_2Fe_{14}B$ and is characterized by anomalous change in the residual magnetization.

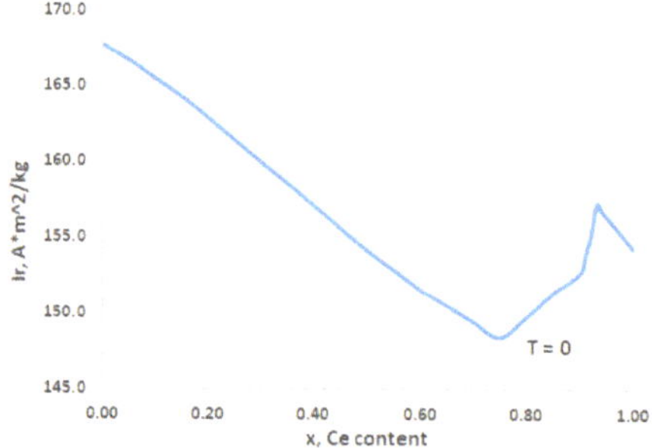

Figure 4. Compositional dependence of the residual magnetization I_r for the $(Nd_{1-x}Ce_x)_2Fe_{14}B$ compounds at T = 0 K.

Figure 5 shows temperature dependences of the coercive force H_c of the $(Nd_{1-x}Ce_x)_2Fe_{14}B$ compounds with x = 0–1.00.

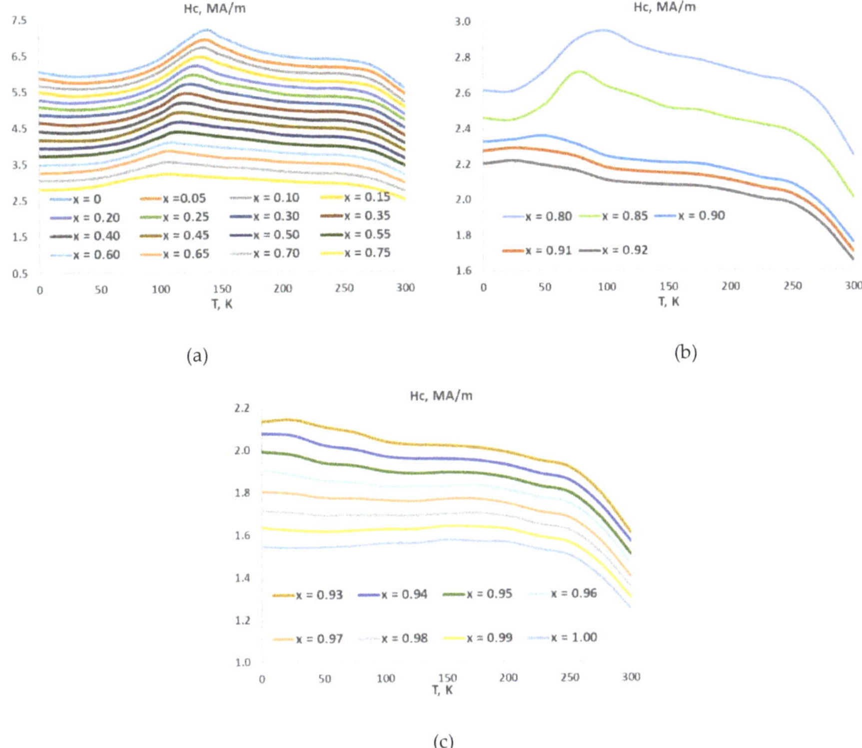

Figure 5. Temperature dependences of the coercive force H_c of the $(Nd_{1-x}Ce_x)_2Fe_{14}B$ compounds with different *x*: (**a**) 0, 0.05, 0.10, 0.15, 0.20, 0.25, 0.30, 0.35, 0.40, 0.45, 0.50, 0.55, 0.60, 0.65, 0.70, 0.75; (**b**) 0.80, 0.85, 0.90, 0.91, 0.92; and (**c**) 0.93, 0.94, 0.95, 0.96, 0.97, 0.98, 0.99, 1.00.

It is seen that, for each Ce content x = 0–0.75, the coercive force H_c reaches the maximum value in the range of $T_{sr}(x)$ (Figure 5a). The monotonic shift of the H_c maximum to the low-temperature range from $H_c(T = 135$ K, $x = 0)$ =7.2 MA/m to $H_c(T = 102$ K, $x = 0.75)$ = 3.2 MA/m correlates with a similar shift of the maximum of normalized $K_2/|K_1|$ ratio, shift of the maximum of I_r, and a decrease in $T_{sr}(x)$ with increasing cerium concentration x (Figures 1 and 3). The maximum in these dependences is related to the competition of two physical phenomena, such as the monotonic increase in the magnetic anisotropy field H_A of $R_2Fe_{14}B$ intermetallics (where R = Nd or Ce) with decreasing temperature and the spin-reorientation below T_{sr}, that facilitates the magnetization reversal process in an external magnetic field.

As is seen, for each Ce content x = 0.80–0.92 (Figure 5b), the coercive force H_c reaches the maximum value in a temperature range of $T_{sr}(x)$. The monotonic shift of the H_c maximum to the low-temperature range from $H_c(T = 100$ K, $x = 0.80)$ =2.9 MA/m to $H_c(T = 25$ K, $x = 0.92)$ = 2.2 MA/m correlates with a similar shift of the maximum of the normalized $K_2/|K_1|$ ratio, a shift of the maximum of I_R, and a decrease in $T_{sr}(x)$ with increasing cerium concentration x (Figures 1–3).

For the compositions with x = 0.93–0.94 (Figure 5c), the maximum of H_c progressively shifts from 25 to ~10 K. For $x > 0.94$, the maximum shifts to 0 K and degenerates. This effect also correlates with a similar shift of the normalized $K_2/|K_1|$ ratio, shift of the maximum of I_r, and a decrease in $T_{sr}(x)$ with increasing cerium concentration x (Figures 1 and 3).

4. Conclusions

The following are our conclusions, which are inferred based on the data obtained in simulating the magnetization reversal process of the $(Nd_{1-x}Ce_x)_2Fe_{14}B$ compounds with x = 0–1.00 and their hysteresis loops in a wide temperature range of 300 to 0 K.

1. The model and algorithm developed for simulating the hysteresis loops of ferromagnets allow us to obtain important data on the effect of uniaxial tetragonal lattice, magnetic anisotropy parameters, and temperature dependences of the magnetic anisotropy constants on the characteristics of hysteresis loops (anisotropy field, residual magnetization, the shape of the hysteresis loop, and coercive force) in a wide temperature range.

2. Results of the simulation indicate that alloying of $Nd_2Fe_{14}B$ with cerium to its contents $x = 0.94$ (1) does not lead to the complete elimination of the negative effect of spin-reorientation phase transition on the residual magnetization of the $(Nd_{1-x}\ Ce_x)_2Fe_{14}B$ compounds and (2) slightly decreases the slope of the magnetization curve and almost does not lead to the improvement of the squareness of the back of hysteresis loop.

3. The alloying of $(Nd_{1-x}Ce_x)_2Fe_{14}B$ with cerium to its contents $x = 0.94$ does not allow the temperature stability of hysteretic characteristics of $(Nd_{1-x}Ce_x)_2Fe_{14}B$-based permanent magnets to be increased in order to ensure their operation at low temperatures without losing magnetic properties.

Author Contributions: Conceptualization, N.B.K. and A.G.S.; methodology and calculations M.V.Z.; formal analysis, Y.S.K. and J.Ć.; data curation, V.P.M.; visualization, N.A.D.; writing—original draft preparation, N.B.K. and M.V.Z.; writing—review and editing, N.B.K.; writing, K.S., M.K., and P.A.P.; project administration, G.S.B. All authors have read and agreed to the published version of the manuscript.

Funding: This study was supported by the Ministry of Science and Higher Education of the Russian Federation, agreement no. 05.604.21.0244, unique identification no RFMEFI60419X0244.

Acknowledgments: The authors thank Associate V.L. Stolyarov (National University of Science and Technology MISiS) for his assistance in the program Hysteresis.

Conflicts of Interest: The authors declare no conflict of interest.

References

1. Li, Z.; Liu, W.; Zha, S.; Li, Y.; Wang, Y.; Zhang, D.; Yue, M.; Zhang, J.; Huang, X. Effects of Ce substitution on the microstructures and intrinsic magnetic properties of Nd–Fe–B alloy. *J. Mag. Mag. Mater.* **2015**, *393*, 551–554. [CrossRef]
2. Zhao, L.; Li, C.; Hao, Z.; Liu, X.; Liao, X.; Zhang, J.; Su, K.; Li, L.; Yu, H.; Greneche, J.-M.; et al. Influences of element segregation on the magnetic properties in nanocrystalline Nd-Ce-Fe-B alloys. *Mater. Charact.* **2019**, *148*, 208–213. [CrossRef]
3. Raghavan, V. The B-Ce-Fe (Boron-Cerium-Iron) System. In *Phase Diagrams of Ternary Iron Alloys*; Indian Institute of Metals: Kolkata, India, 1992; Volume 6A, pp. 297–300.
4. Herbst, J.F. $R_2Fe_{14}B$ materials: Intrinsic properties and technological aspects. *Rev. Mod. Phys.* **1991**, *63*, 819–898. [CrossRef]
5. Hirosawa, S.; Matsuura, Y.; Yamamoto, H.; Fujimura, S.; Sagawa, M.; Sagawa, M.; Yamauchi, H. Magnetization and magnetic anisotropy of $R_2Fe_{14}B$ measured on single crystals. *J. Appl. Phys.* **1986**, *59*, 873–879. [CrossRef]
6. Fruchart, R.; L'Heritier, P.; Dalmas de Reotier, P.; Fruchart, D.; Wolfers, P.; Coey, J.M.D.; Ferreira, L.P.; Guillen, R.; Vulliet, P.; Yaouanc, A. Mossbauer spectroscopy of $R_2Fe_{14}B$. *J. Phys. F Met. Phys.* **1987**, *17*, 483–502. [CrossRef]
7. Grössinger, R.; Sun, X.K.; Eibler, R.; Buschow, K.H.J.; Kirchmayr, H.R. Temperature dependences of anisotropy fields and initial susceptibilities in $R_2Fe_{14}B$ compounds. *J. Magn. Magn. Mater.* **1986**, *58*, 55–60. [CrossRef]
8. Haskel, D.; Lang, J.C.; Islam, Z.; Cady, A.; Srajer, G.; van Veenendaal, M.; Canfield, P.C. Atomic Origin of Magnetocrystalline Anisotropy in $Nd_2Fe_{14}B$. *Phys. Rev. Lett.* **2005**, *95*, 217207. [CrossRef]
9. Colin, C.V.; Ito, M.; Yano, M.; Dempsey, N.M.; Suard, E.; Givord, D. Solid-solution stability and preferential site occupancy in $(R\text{-}R')_2Fe_{14}B$ compounds. *Appl. Phys. Lett.* **2016**, *108*, 242415. [CrossRef]
10. Bolzoni, F.; Gavigan, J.P.; Givord, D.; Li, H.S.; Moze, O.; Pareti, L. 3d magnetism in $R_2Fe_{14}B$ compounds. *J. Magn. Magn. Mater.* **1987**, *66*, 158. [CrossRef]

11. Alam, A.; Khan, M.; McCallum, R.W.; Johnson, D.D. Site-preference and valency for rare-earth sites in (R-Ce)$_2$Fe$_{14}$B magnets. *Appl. Phys. Lett.* **2013**, *102*, 042402. [CrossRef]
12. Abache, C.; Oesterreicher, J. Magnetic anisotropies and spin reorientations of R$_2$Fe$_{14}$B-type compounds. *J. Appl. Phys.* **1986**, *60*, 3671. [CrossRef]
13. Hallemans, B.; Wollants, P.; Roos, J.R. Thermodynamic assessment of the Fe-Nd-B phase diagram. *J. Phase Equilib.* **1995**, *16*, 137–149. [CrossRef]
14. Li, Z.B.; Shen, B.G.; Zhang, M.; Hu, F.X.; Sun, J.R. Substitution of Ce for Nd in preparing R$_2$Fe$_{14}$B nanocrystalline magnets. *J. Alloy Compd.* **2015**, *628*, 325–328. [CrossRef]
15. Wang, J.; Liang, L.; Zhang, L.T.; Yano, M.; Terashima, K.; Kada, H.; Kato, S.; Kadono, T.; Imada, S.; Nakamura, T.; et al. Mixed-valence state of Ce and its individual atomic moments in Ce$_2$Fe$_{14}$B studied by soft X-ray magnetic circular dichroism. *Intermetallics* **2016**, *69*, 42–46. [CrossRef]
16. Fan, X.; Ding, G.; Chen, K.; Guo, S.; You, C.; Chen, R.; Lee, D.; Yan, A. Whole process metallurgical behavior of the high-abundance rare-earth elements LRE (La, Ce and Y) and the magnetic performance of Nd$_{0.75}$LRE$_{0.25}$-Fe-B sintered magnets. *Acta Mater.* **2018**, *154*, 343–354. [CrossRef]
17. Jiang, Q.; Zhong, Z. Research and development of Ce-containing Nd$_2$Fe$_{14}$B-type alloys and permanent magnetic materials. *J. Mater. Sci. Technol.* **2017**, *33*, 1087–1096. [CrossRef]
18. Pei, K.; Zhang, X.; Lin, M.; Yan, A.R. Effects of Ce-substitution on magnetic properties and microstructure of Nd–Pr–Fe–B melt-spun powders. *J. Magn. Magn. Mater.* **2016**, *398*, 96–100. [CrossRef]
19. Yan, C.; Guo, S.; Chen, R.; Lee, D.; Yan, A. Effect of Ce on the Magnetic Properties and Microstructure of Sintered Didymium-Fe-B Magnets. *IEEE Trans. Magn.* **2014**, *50*, 2102605. [CrossRef]
20. Pathak, A.K.; Gschneidner, K.A.; Khan, M.; McCallum, R.W.; Pecharsky, V.K. High performance Nd-Fe-B permanent magnets without critical elements. *J. Alloys Compd.* **2016**, *668*, 80–86. [CrossRef]
21. Fuerst, C.D.; Herbst, J.F.; Sarrao, J.L.; Migliori, A. Resonant ultrasound measurements of elastic constants of melt-spun of R$_2$Fe$_{14}$B compounds (R=Ce, Pr, Nd, Er). *J. Appl. Phys.* **1994**, *75*, 6625–6627. [CrossRef]
22. Susner, M.A.; Conner, B.S.; Saparov, B.I.; McGuire, M.A.; Crumlin, E.J.; Veith, G.M.; Cao, H.; Shanavas, K.V.; Parker, D.S.; Chakoumakos, B.C.; et al. Flux growth and characterization of Ce-substituted Nd$_2$Fe$_{14}$B single crystals. *J. Mag. Mag. Mater.* **2017**, *434*, 1–9. [CrossRef]
23. Menushenkov, V.; Savchenko, A.; Stoliarov, V.; Zheleznyi, M.; Kolchugina, N.; Dormidontov, N.; Skotnicova, K.; Koshkidko, Y. Simulation of demagnetization reversal and hysteretic loops of (Nd, Pr)$_2$Fe$_{14}$B magnets at low temperatures. In *Proceedings of 28-th International Conference on Metallurgy and Materials, Metal 2019, Brno, Czech Republic, 22–24 May 2019*; Tanger Ltd.: Ostrava, Czech Republic, 2019; pp. 1621–1627.
24. Jiang, S.; Li, H.; Xu, J.; Gao, R. Research on the origin of magneto-crystalline anisotropy in R$_2$Fe$_{14}$B compounds. *J. Mag. Mag. Mater.* **1994**, *136*, 294–300.
25. Xu, W.G.; Liu, Y.L.; Ji, S.Q.; Chu, D.P.; Xu, Y.; Yang, G.L. Temperature dependence of the anisotropy constants K$_1$ and K$_2$ of R$_2$Fe$_{14}$B (R. = Ce, Pr, Gd). *Acta Sin.* **1986**, *35*, 1592–1597.

Article

Modeling the Maximum Magnetic Entropy Change of Doped Manganite Using a Grid Search-Based Extreme Learning Machine and Hybrid Gravitational Search-Based Support Vector Regression

Sami M. Ibn Shamsah [1] **and Taoreed O. Owolabi** [2,*]

1 Department of Mechanical Engineering, College of Engineering, University of Hafr Al Batin, P.O. Box 1803, Hafr Al Batin 31991, Saudi Arabia; sibnshamsah@uhb.edu.sa

2 Physics and Electronics Department, Adekunle Ajasin University, Akungba Akoko 342111, Ondo State, Nigeria

* Correspondence: taoreed.owolabi@aaua.edu.ng

Received: 1 April 2020; Accepted: 14 April 2020; Published: 16 April 2020

Abstract: The thermal response of a magnetic solid to an applied magnetic field constitutes magnetocaloric effect. The maximum magnetic entropy change (MMEC) is one of the quantitative parameters characterizing this effect, while the magnetic solids exhibiting magnetocaloric effect have great potential in magnetic refrigeration technology as they offer a green solution to the known pollutant-based refrigerants. In order to determine the MMEC of doped manganite and the influence of dopants on the magnetocaloric effect of doped manganite compounds, this work developed a grid search (GS)-based extreme learning machine (ELM) and hybrid gravitational search algorithm (GSA)-based support vector regression (SVR) for estimating the MMEC of doped manganite compounds using ionic radii and crystal lattice parameters as descriptors. Based on the root-mean-square error (RMSE), the developed GSA-SVR-radii model performs better than the existing genetic algorithm (GA)-SVR-ionic model in the literature by 27.09%, while the developed GSA-SVR-crystal model performs better than the existing GA-SVR-lattice model in the literature by 38.34%. Similarly, the developed ELM-GS-crystal model performs better than the existing GA-SVR-ionic model with a performance enhancement of 14.39% and 20.65% using the mean absolute error (MAE) and RMSE, respectively, as performance measuring parameters. The developed models also perform better than the existing models using correlation coefficient as the performance measuring parameter when validated with experimentally measured MMEC. The superior performance of the present models coupled with easy accessibility of the descriptors definitely will facilitate the synthesis of doped manganite compounds with a high magnetocaloric effect without experimental stress.

Keywords: magnetocaloric effect; support vector regression; extreme learning machine; maximum magnetic entropy change; gravitational search algorithm

---

1. Introduction

Magnetocaloric compounds are of technological and scientific interest mainly because of their significance in magnetic refrigeration, which has shown high potential in replacing the conventional compression–expansion cycle of a gas cooling system [1–4]. These compounds are significant in magnetic refrigeration technology since they serve as the refrigerants and the major component of the cooling system characterized with lower noise, a high degree of compactness, high cooling efficiency and environmental friendliness as they do not require ozone-depleting gases [5,6]. Before the discovery of manganite-based compounds as a potential source of a high magnetocaloric effect (which is a physical phenomenon that measures the change in internal temperature of a compound when subjected to adiabatic magnetization), gadolinium was among the well-explored elements with significant magnetocaloric effect near room temperature [7,8]. However, magnetocaloric effect can be observed in many compounds, while the application of these compounds in magnetic refrigeration technology is hindered by the appearance of the magnetocaloric effect at relatively high magnetic fields and/or at transition temperatures differing greatly from room temperature. Manganite-based compounds combine many unique features that make them fit well into magnetic refrigeration technology [9]. They demonstrate a high magnetocaloric effect at low applied magnetic fields, relatively cheap elemental compositions, high stability (especially in some corrosive environments) and physical parameters that can be easily tuned through doping mechanisms [5,10,11]. Doping concentration plays a crucial role in altering the magnetocaloric effect as well as the maximum magnetic entropy change (MMEC) of manganite-based compounds, since A-site average ionic radius, ratio of Mn^{4+} and Mn^{3+} ions and A-site disorder are strongly affected by the nature as well as the concentration of dopants [12–14]. A-site average ionic radius contributes mainly to magnetic properties of doped manganite through an energy bandwidth of electron alteration that is related to the observed double-exchange interaction, while A-site disorder measures A-site ionic distribution variance [15,16]. The present work developed models through which the maximum magnetic entropy change (MMEC) of manganite compounds could be influenced by the introduction of dopants into the parent manganite compound.

The observed physical properties in manganite-based compounds come from strong competition among antiferromagnetic superexchange interactions, ferromagnetic double-exchange interaction and spin–phonon coupling [17,18]. Zener's double-exchange theory explains the importance of Mn^{4+} and Mn^{3+} ions to the ferromagnetic feature of these compounds, while the exhibition of magnetic properties is well captured within electron–phonon interaction described in Jahn–Teller formalism [4]. Doping of sites in the parent manganite compound with divalent alkaline–earth ions, monovalent alkaline or any other transition metals induces distortion in the lattice and changes the Mn^{4+} and Mn^{3+} ion network that, in turn, alters the magnetocaloric effect as well as the maximum magnetic entropy change (MMEC) of manganite-based compounds. Identifying the relationship between the nature as well as the concentration of the dopants supports the understanding of their physical properties, while the synthesis of new manganite-based compounds having multifunctional applications becomes possible. Some screening approaches to developing and identifying materials with a large magnetocaloric effect have been proposed in the literature [19], [20]. However, machine learning techniques have been identified as effective and efficient tools for determining the influence of dopants on the physical properties (such as magnetic ordering temperature, relative cooling power and magnetocaloric effect) of doped manganite-based compounds [6,21–25]. The proposed grid search-based extreme learning machine (GS-ELM) and hybrid gravitational search-based support vector regression model the relationship between lattice distortion as well as elemental compositions on the MMEC when dopants are incorporated into the crystal structure of the parent manganite-based compounds.

Support vector regression (SVR) is a type of machine learning technique developed using statistical learning theory [26]. It acquires support vectors linking the descriptors with the desired target at the training phase while the acquired vectors are further validated. The algorithm maintains a high level of precision and accuracy while dealing with many real-life as well as complex problems [27–30]. The robustness and effectiveness of an SVR-based model have been attributed to its non-convergence to local minima, sound mathematical background and proper tuning of its user-defined hyperparameters [31,32]. As such, the SVR algorithm and its hybrid have been extensively applied to many problems in science and engineering. A gravitational search algorithm (GSA) was used in this research work for SVR hyperparameter optimization. GSA is a population-based optimization algorithm that is capable of navigating potential solutions in the search space using Newtonian mechanical principles [33]. Its fast convergence to a global solution, avoidance of premature convergence and dependence of its convergence on a limited number of parameters contribute to its uniqueness that required its implementation in this work.

The extreme learning machine (ELM) is a computational intelligence algorithm in which a single-hidden-layer feedforward network is trained in a unique way that ensures a very fast speed together with significant computational scalability [34–36]. Three different layers are associated with the ELM; these layers include the input layer, the hidden or latent layer and the output layer. The parameters of the hidden layer are generated randomly and kept unchanged throughout the training phase of model development. Another stronghold of the ELM algorithm is the implementation of the least-square method for solving the optimization problem instead of obtaining partial derivatives of model parameters through the chain rule; thus, output weights are obtained analytically [37]. With the stochastic latent layer weights, the ELM still maintains universal approximation strength in acquiring the relationship between descriptors and targets. These unique capacities of the ELM algorithm are explored in the present work for modeling the MMEC of doped manganite compounds using lattice parameters and ionic radii as descriptors.

The results of modeling and simulation of the MMEC of doped manganite compounds show that the developed GSA-SVR-radii model, which employs the ionic radii and the concentrations of dopants as descriptors, performs better than the existing genetic algorithm(GA)-SVR-ionic model [38] in the literature by 27.09% on the basis of the root-mean-square error (RMSE), while the developed GSA-SVR-crystal model, which utilizes crystal lattice parameters of the compound, performs better than the existing GA-SVR-lattice model [38] in the literature by 38.34% using RMSE as a performance measuring parameter. The superior performance demonstrated by the proposed model definitely helps identify the manganite compounds with a high magnetocaloric effect and ultimately hastens the practical implementation of magnetic refrigeration technology.

The remaining part of this manuscript is organized as follows: Section 2 describes the mathematical formulation of the support vector regression algorithm, gravitational search algorithm and extreme learning machine. Section 3 discusses the computational strategies employed for model development. The description of the dataset used for modeling and simulation is also presented in the section. Section 4 presents the results and compares the present work with the existing models. Section 5 concludes the manuscript.

2. Mathematical Formulation of the Proposed Models

The mathematical background and formulation of the proposed techniques are presented in this section. The algorithms described include support vector regression, gravitational search algorithm and extreme learning machine.

2.1. Description of the Support Vector Regression Algorithm

The support vector regression algorithm is a learning computational algorithm that acquires useful information and intricacies linking the descriptors with the desired target so as to ultimately attain excellent generalization [39,40]. The algorithm was first developed using the structural risk minimization inductive principle that aims at minimizing Vapnik–Chervonenkis dimension and empirical risk simultaneously. The algorithm acquires relevant and highly informative patterns from the training set of data $(a_1, MMEC_1)..........(a_j, MMEC_j) \in \Re^{N} X\Re$ and constructs a regression model that governs future predictions. The regression equation can be generalized as presented in Equation (1):

$$MMEC(a) = \langle \omega, a \rangle + b, \omega \in N, b \in \Re \tag{1}$$

where $MMEC$ is the estimated maximum magnetic entropy change; a_j represents the descriptors, which are the crystal lattice parameters and the applied magnetics field for the GSA-SVR-crystal model, whereas it represents the concentration of the dopants, ionic radii and the applied magnetic field for the GSA-SVR-ionic model; $\langle .,. \rangle$ is the dot product and N is the input pattern space.

Prior to the construction of the regression function in either the ordinary or the high dimensional feature space, the optimization problem depicted in Equation (2) is solved with a special consideration given to the constraints in Equation (3) [41–43]:

$$\frac{1}{2}||\omega||^2 + C\sum_{j}^{m}\left(\xi_j^* + \xi_j\right) \tag{2}$$

$$\begin{cases} MMEC_j - \langle\omega, a\rangle - b \leq \varepsilon + \xi_j \\ \langle\omega, a\rangle - MMEC^{\exp}{}_j + b \leq \varepsilon + \xi_j^* \\ \xi_j^*, \xi_j \geq 0 \end{cases} \tag{3}$$

where $||.||$ represents the Euclidian norm, $MMEC^{\exp}{}_j$ stands for the measured maximum magnetic entropy change, ξ_j^*, ξ_j are the slack variables and ε is the maximum allowable deviation of all training data points from the measured values.

Deviations of data points from the measured values are penalized or regularized in the SVR algorithm. The parameter C in Equation (2) measures the degree of penalty and is referred to as the penalty factor. In addition, the slack variables ξ_j^*, ξ_j are introduced into the constraint equation presented in Equation (3) purposely to eradicate or minimize the possibility of data points falling outside $\varepsilon - tube$. Mapping of data points that could not be solved in ordinary space to the high dimensional feature space was done in the present work using the Gaussian kernel function presented in Equation (4):

$$\chi(a_j, a_i) = \exp\left(\frac{-1}{2}\left\{\frac{||a_j - a_i||^2}{\sigma}\right\}\right) \tag{4}$$

where σ represents the kernel option.

The final regression equation after transformation is presented in Equation (5):

$$MMEC(a, \lambda) = \sum_{j=1}^{m}\left(\lambda_j^* - \lambda_j\right)\chi(a_j, a_i) + b \tag{5}$$

where λ^* and λ are the Lagrange multipliers.

The performance of the SVR-based model is strongly influenced by the model hyperparameters which include C, σ and ε [44–46]. Therefore, in order to develop a robust model that is characterized with a high degree of precision, these hyperparameters must be tuned to optimum values. The hyperparameters are tuned in this work using the gravitational search algorithm.

2.2. Physical Principles of the Gravitational Search Algorithm

The gravitational search algorithm (GSA) is a class of heuristic population-based algorithms employed for solving optimization problems [33]. Its fundamental principle of navigating through a search space of possible solutions is Newtonian mechanical formalism. In this approach, the solution search space is explored and exploited by the agents, which are known as objects in Newtonian description. The interaction of these agents with one another is controlled by the value of their masses as well as gravitational pull [27,47,48]. An agent with heavy mass corresponds to a good or global solution and moves very slowly in the solution space, whereas agents of lighter masses are attracted towards heavy agents. A stepwise description of the algorithm operational principles goes as follows:

Step 1: Agent initialization. For q-number of initial population of agents, initialize the position of the agents randomly. Evaluate the fitness of each of the agents using a defined objective function. The fitness of the agents in this work was evaluated using the root-mean-square error between the measured MMEC and predicted values obtained from objective function implementation. $Y_k = (y_k^1 \ldots . y_k^D, \ldots . y_k^P)$ specifies the position of kth agent where $k = 1, 2, \ldots \ldots, K$.

Step II: Calculation of agents' masses. Using the minimum and maximum values of the fitness of the entire population, the mass of each of the agents is computed at the *ith* iteration through the implementation of Equations (6) and (7):

$$P_k = \frac{fitness_k(i) - Max\_fitness(i)}{Min\_fitness(i) - Max\_fitness(i)} \tag{6}$$

$$M_k(i) = \frac{P_j(i)}{\sum_{k=1}^{K} P_j(i)} \tag{7}$$

where $fitness_k(i), Min\_fitness(i)$ and $Max\_fitness(i)$ represent the fitness of the *kth* agent, the minimum fitness of the entire population and the maximum fitness of the entire population, respectively.

Step III: Computation of Newtonian gravitational force. The Newtonian gravitational attraction $F_k^z(i)$ between the *kth* and *zth* agents is depicted in Equation (8), where $F_{kz}^D(i)$ and $r_{kz}(i)$ are defined, respectively, in Equations (9) and (10) [49]:

$$F_k^z(i) = \sum_{k=1, k\neq z}^{K} rand_z F_{kz}^D(i) \tag{8}$$

$$F_{kz}^D(i) = G(i)\frac{M_k(i)M_{az}(i)}{r_{kz}(i)+\delta}\left(y_k^D(i) - y_z^D(i)\right) \tag{9}$$

$$r_{kz}(i) = \left\|Y_k(i), Y_z(i)\right\|_2 \tag{10}$$

where $r_{kz}(i)$, $G(i)$ and $rand_z$ stand for the Euclidian distance between the k and z agents, the gravitational constant defined as $G_0 \exp\left(-\alpha\left(\frac{i}{\tau}\right)\right)$ and the random number spanning in the range of 0 to 1, respectively. Similarly, $M_{az}(i)$, τ and δ represent the active gravitational mass of the *zth* agent, the maximum iteration and a small constant value, respectively.

Step IV: Calculation of the acceleration with which each of the agents is navigating. The acceleration of the agents is calculated using Equation (11):

$$a_k^D(i) = \frac{F_k^D(i)}{M_k(i)} \tag{11}$$

Step V: Calculation of the position and velocity of the agents. Equations (12) and (13) present the expression for calculating the position and velocity of the agents in the population, respectively.

$$y_k^D(i+1) = y_k^D(i) + v_k^D(i+1) \tag{12}$$

$$v_k^D(i+1) = rand_k x\ v_k^D(i) + a_k^D(i) \tag{13}$$

Step VI: Stopping criteria of the algorithm. Steps I to V are repeated until the maximum number of iterations is reached.

2.3. Mathematical Background of the Extreme Learning Machine

The extreme learning machine is a novel learning algorithm of training a single-hidden-layer feedforward neural network [34,50]. The algorithm analytically determines the output weights using the Moore–Penrose generalized inverse method, while the input weights as well as the hidden biases are randomly chosen. Suppose a single-hidden-layer feedforward neural network is to be trained with the aid of *t* number of hidden neurons coupled with the nonlinear activation function $g(x)$ and *j* number of training samples $(\mathbf{x}_i, \mathbf{t}_i)$ so that $\mathbf{x}_i = [x_{i\,1}, \ldots\ldots, x_{ip}]^T \in \mathbf{R}^p$ and $MMEC_i = [MMEC_{i\,1}, \ldots\ldots, MMEC_{i\,m}]^T \in \mathbf{R}^m$ (where *p* and *m* represent the number of data points for output and input layers, respectively).

The operational principle of the ELM algorithm transforms the nonlinear system to a linear system using the transformation relation presented in Equation (14) [51].

$$\mathbf{H}\beta = \mathbf{T} \tag{14}$$

where $\mathbf{H} = \{h_{ip}\}(i = 1,\ldots\ldots,k \text{ and } p = 1,\ldots.,t)$, pth is the hidden neuron output with respect to $\mathbf{x}_i = h_{ip} = f(\mathbf{w}_p.\mathbf{x}_i + b_p)$, $\mathbf{w}_p = [w_{p\,1},\ldots\ldots., w_{p\,t}]^T$ = weight vector, linking the pth hidden neuron to the input neurons, b_p = bias of the pth hidden neuron, $\beta = [\beta_1,\ldots\ldots., \beta_i]^T$, the output weight matrix $\beta_p = [\beta_{p1},\ldots\ldots., \beta_{p\,m}]^T (p = 1,\ldots,i)$, the weight vector linking the pth hidden neuron to the output neurons and $\mathbf{T} = [t_1,\ldots\ldots., t_k]^T$ is the matrix containing the descriptors.

The value of the MMEC of the doped manganite compound was estimated using the least-square method with the minimum norm relation presented in Equation (15) after the implementation of the linearization process of Equation (14):

$$\hat{\beta} = \mathbf{H}^{\dagger}\mathbf{T} \tag{15}$$

where $\mathbf{H}^{\dagger}$ = Moore–Penrose generalized inverse of matrix $\mathbf{H}$.

3. Dataset Description and Computational Implementation of the Proposed Models

This section describes the computational method employed for the hybridization of the gravitational search algorithm with the support vector regression algorithm. The computational details of the proposed ELM-based model are also presented as well as the description of the dataset. The chemical formula of the manganite compound, whose MMEC the developed models can effectively estimate, is also described and presented.

3.1. Dataset Description and Chemical Formula of the Doped Manganite that Can be Incorporated into the Developed Models

The proposed models in this work were developed using one hundred experimentally measured MMEC of different manganite-based compounds extracted from the literature [7,8,11,15,18,52–64]. The descriptors to the model are two sets of data. The descriptors to the proposed GSA-SVR-crystal model are the lattice distortions (as measured by the crystal lattice parameters) suffered by the manganite crystal structure due to the introduction of dopants into the parent manganite [54,65–72]. The applied magnetic field is also included as a descriptor. The significance of lattice distortion to magnetic properties of the manganite-based compound can be inferred from the influence of lattice distortion that accompanied doping on the Mn^{4+} and Mn^{3+} ion network. Similarly, since ionic radii of the dopants have a strong influence on the magnetic properties of manganite-based compounds, these ionic radii and the concentrations of dopants serve as the descriptors to the developed GSA-SVR-radii model. This model also employs the value of the applied field as a descriptor. The chemical formula that explains the implementation of the proposed GSA-SVR-radii model on manganite-based compounds is presented in Equation (16).

$$R_{1-a-b}A_aB_bMn_{1-c}C_cO_3 \tag{16}$$

where R and A represent the rare-earth element and alkaline-earth element, respectively, while B and C are any dopant of choice. The concentrations of the dopants are represented by the lower case of the dopant symbols. Implementation of the proposed GSA-SVR-radii model helps in determining the nature and the concentration of dopants that give the desired value of the MMEC, and the outcome of the model can be verified experimentally. The GSA-SVR-radii model is robust in the sense that it can easily incorporate four different dopants including the rare-earth elements at various concentrations. This is advantageous as a variety of dopants can be explored as quickly as possible. During model implementation, the dopant that is absent in the chemical structure is assigned a zero value. Suppose that the model is to be implemented to determine the MMEC of the $Pr_{0.6}Ca_{0.1}Sr_{0.3}Mn_{0.8}Fe_{0.2}O_3$

compound; the descriptors are the ionic radii of elements Pr, Ca, Sr and Fe, while a, b and c are 0.1, 0.3 and 0.2, respectively. Table 1 shows the results of statistical analysis conducted on the dataset. This is insightful in determining the content and the nature of the dataset employed for modeling and simulation. The correlation coefficients between the descriptors and the target clearly show an insignificant degree of linear relationship between them. Therefore, the need for nonlinear modeling techniques such as those presented in this work becomes necessary.

Table 1. Statistical analysis of the dataset.

Quantity	Minimum	Maximum	Mean	Standard Deviation	Correlation Coefficient
R(pm)	109.8	143	115.696	4.652	−0.272
A(pm)	0	152	103.83	51.2	0.389
B(pm)	0	152	79.92	66.139	0.334
C(pm)	0	100	23.07	38.523	0.14
a	0	0.45	0.162	0.126	0.261
b	0	0.45	0.161	0.169	0.441
c	0	0.3	0.029	0.048	−0.189
Distortion along *a*-axis	5.429	5.553	5.488	0.033	−0.557
Distortion along *b*-axis	5.438	13.467	6.483	1.307	0.102
Distortion along *c*-axis	5.419	359.95	12.036	35.311	0.105
Applied field (T)	0.05	6	3.483	1.758	0.583
MMEC	0.003	7.14	2.117	1.567	

3.2. Computational Hybridization of the Gravitational Search and Support Vector Regression Algorithms

The computational part of this work was carried out within the MATLAB computing environment. The available data points for modeling and simulation were partitioned into training and testing sets after being randomized to ensure uniformity and prevent uneven distribution of the data points. The hyperparameters of the SVR algorithm optimized using the GSA include the regularization factor, epsilon and kernel option of the best kernel function, while the hyperparameter that controls the hyperplanes was maintained as 1E−7. Step-by-step procedures for the computational hybridization of GSA with the SVR algorithm are detailed as follows:

Step I: *Data division.* The 100 available data points were divided into training and testing sets in the ratio of 8:2. The descriptors in each of the partitions consisted of nine different descriptors for the GSA-SVR-radii model, while four descriptors were attributed to the GSA-SVR-crystal model. The developed GSA-SVR-radii model employs the ionic radii of the dopants as well as their concentrations, while the developed GSA-SVR-crystal model utilizes the crystal lattice distortion of the manganite-based compound due to the introduction of dopants into the crystal lattice structure of the parent manganite.

Step II: *Hyperparameter initialization.* The agents were populated within the Newtonian description of the gravitational search algorithm. Each agent encodes the regularization factor, epsilon and kernel option of a chosen function. For K-number of agents in a search space, each agent position is represented as $Y_k = (y_k^1.....y_k^D, ...y_k^P)$ where $k = 1, 2, ..., K$ and y_k^D represents the current agent position in the Dth dimension. In the present work, the dimension is three since there are three parameters to be optimized.

Step III: *Rating of fitness of each of the agents in the population.* The fitness of each of the agents was determined using the root-mean-square error (RMSE) between the model estimates and the measured values of the MMEC. The agent in the testing set of the data, which is characterized by the lowest RMSE, has the best fitness, while the agent with the highest RMSE becomes the worst agent in the population.

Step IV: *Inertial mass of the agents in the population and gravitational pull computation.* The inertial mass is computed using Equations (6) and (7). The Newtonian gravitation pull is computed using Equation (8).

Step V: *Computation of acceleration, position and velocity of the agents in the population.* The acceleration, position and velocity of each of the agents in the population were computed using Equations (11)–(13), respectively.

Step VI: *Stopping condition*. After each iteration, Steps II to V are repeated, while the position of each of the agents in the population is updated until the algorithm attains a global solution. The global solution is indicated by a zero value of RMSE or the same value of RMSE for fifty consecutive numbers of iteration.

3.3. *Computational Implementation of the Grid Search-Based Extreme Learning Machine*

The extreme learning machine is an efficient and robust algorithm that adequately trains a single-hidden-layer feedforward network and ultimately achieves a high degree of precision as well as reduced computational time. The ELM algorithm has one parameter (i.e., the number of hidden nodes) and a nonlinear function that must be tuned and optimized in order to enhance the precision of the algorithm. This parameter and the function are optimized in this work using a grid search (GS) approach. The number of hidden nodes was sought within a specified range for each of the available activation functions. Before the computational implementation of the ELM-GS model, the dataset was partitioned and randomized in a similar manner to the approach used in developing the GSA-SVR models. In this case also, two different models were developed. The developed ELM-GS-radii model uses ionic radii and concentrations of dopants as descriptors similar to the descriptors of the GSA-SVR-radii model, whereas the developed ELM-GS-crystal model employs lattice parameters as descriptors. Of course, all the developed models in this work include the applied magnetic field during magnetization measurement to the descriptors. The details of the computational strategies employed in developing the two ELM-GS-based models are described step by step as follows:

Step I: *Data randomization and partition*. The datasets used for developing the GSA-SVR-based models were also employed here. This allows a fair comparison between the generalization and predictive strength of the developed models.

Step II: *Generation of input weights and biases using pseudorandom number generator*. Using the pseudorandom number generator (seeding) in the MATLAB computing environment, input weights $\mathbf{w}_p$ and biases b_p were generated. With the Mersenne Twister generator using seed N [rng(N)], these weights and biases were optimized by searching for values of N between 0 to 100 that result in the lowest RMSE between the measured and estimated MMEC. The optimum initial weights and biases were saved and implemented for the rest of the modeling and simulation stages.

Step III: *Computation of hidden-layer output matrix* **H**. For the selected activation function from the pool of available functions (Sine (sin), sigmoidal (sig), radial basis (radbas), hardlim (hardlim) and triangular basis (tribas) functions), compute the hidden-layer output weights for each of the hidden nodes, which spans between (1,100) using the training set of data.

Step IV: *Determination of output weights* $\hat{\beta}$ *and computation of the MMEC*. Compute the output weights and calculate the MMEC for each of the manganite-based compounds using the Moore–Penrose generalized inverse matrix for each of the hidden nodes in Step III.

Step V: *Selection of the best model for each of the activation functions*. Using the obtained input weights, biases and output weights for each of the number of nodes, determine the RMSE between the measured MMEC and the estimated values and select the model with the lowest RMSE using the testing set of data. The number of nodes and the activation function corresponding to the best model are also saved.

Step VI: *Selection of the overall best model*. Repeat Steps III to V and select the overall best model. The input weights, biases, output weights, activation function and the number of nodes corresponding to the best models are saved for future implementation.

4. Results and Discussion

This section discusses the outcome of the research work. The dependence of the convergence of the GSA-SVR-based model on the initial number of agents exploiting and exploring the population is discussed and presented. The comparison between the results of this work and the existing models is also presented in this section.

4.1. Searching for the Optimum Hyperparameters of the Developed Models

The dependence of the convergence of the developed GSA-SVR-radii and GSA-SVR-crystal models to the initial number of agents exploring and exploiting the search space are presented in Figures 1 and 2, respectively. For the convergence of the GSA-SVR–radii model presented in Figure 1, the convergence to local minima was observed when ten, thirty and fifty agents explored the search space, while the model converged to a global solution when the number of agents was increased to seventy. Above this value, no further minimization of the objective function was possible. Similarly, Figure 2 presents the convergence for the GSA-SVR-crystal model that employs the crystal lattice parameters and the applied magnetic field as descriptors. The developed GSA-SVR-crystal model converged to a local solution when ten agents were exploring and exploiting the search space. The global solution was reached by thirty agents, while the increase in the number of initial agents returned the model to a local solution convergence. The optimum values of the SVR hyperparameters for the GSA-SVR-radii and GSA-SVR-crystal models are presented in Table 2. The table also presents the optimum activation functions and the number of hidden nodes for the developed ELM-GS-radii and ELM-GS-crystal models.

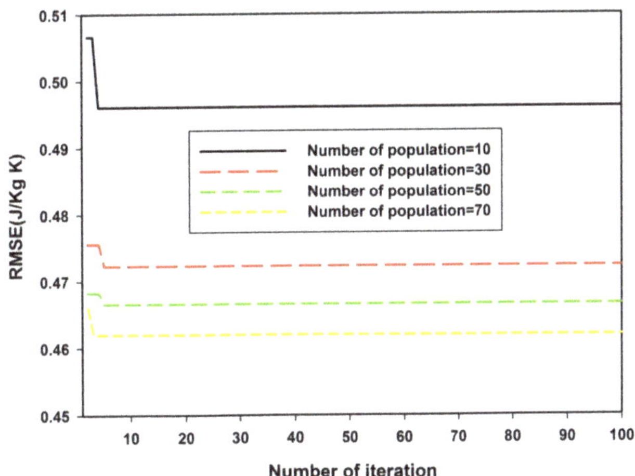

Figure 1. Dependence of convergence of GSA-SVR-radii model on the density of search space.

Table 2. Optimum hyperparameters of support vector regression (SVR)-based models as obtained using a gravitational search algorithm (GSA) and extreme learning machine (ELM)-based models as obtained using grid search (GS).

	GSA-SVR-Radii	GSA-SVR-Crystal	ELM-GS-Radii	ELM-GS-Crystal
Penalty factor	944.3013	592.9928	—	—
Epsilon	0.1003	0.3758	—	—
Kernel option	0.8198	0.2516	—	—
Number in the population	70	30	—	—
Activation function	—	—	Hardlim	Sin
Number of hidden nodes	—	—	58	24

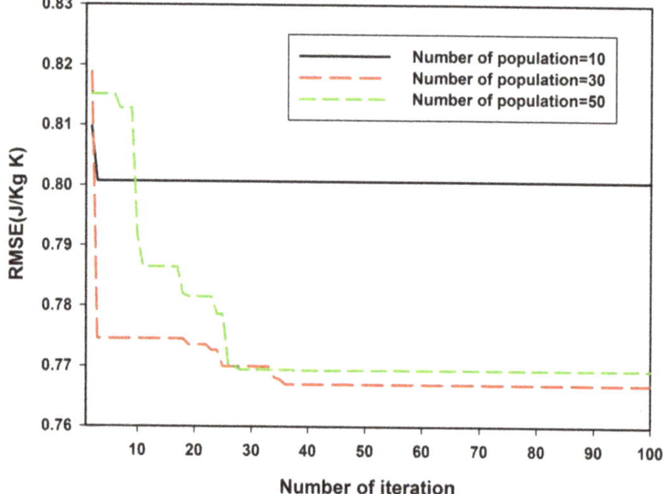

Figure 2. Dependence of convergence of GSA-SVR-crystal model on the density of search space.

4.2. Generalization and Predictive Strength Comparison Between the SVR- and ELM-Based Models

The generalization and predictive strengths of the developed models are compared using three different performance measuring parameters, which include correlation coefficient (CC), mean absolute error (MAE) and root-mean-square error (RMSE). Figures 3 and 4 compare the training and testing stages of the GSA-SVR-radii and GSA-SVR crystal models using CC and MAE performance measuring parameters, respectively.

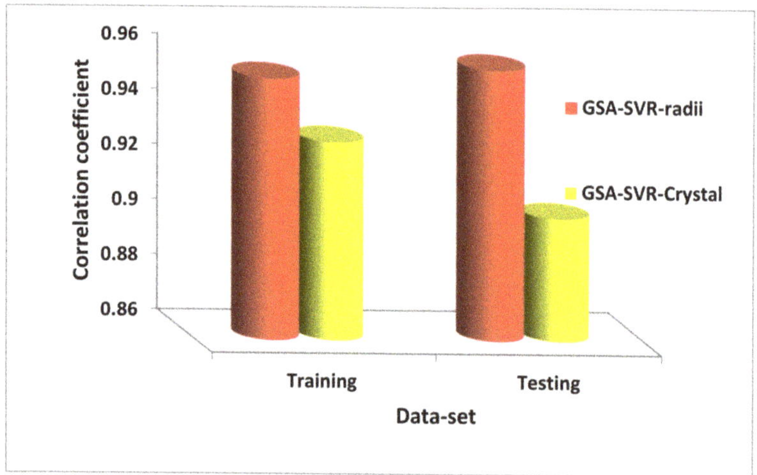

Figure 3. Comparison between the GSA-SVR-radii and GSA-SVR-Crystal models on the basis of the correlation coefficient.

The training stage of the GSA-SVR-radii model demonstrates better performance than that of the GSA-SVR-crystal model with a performance improvement of 2.39%, whereas a percentage enhancement of 5.59% was achieved on the testing set of data using CC as the performance measuring parameter.

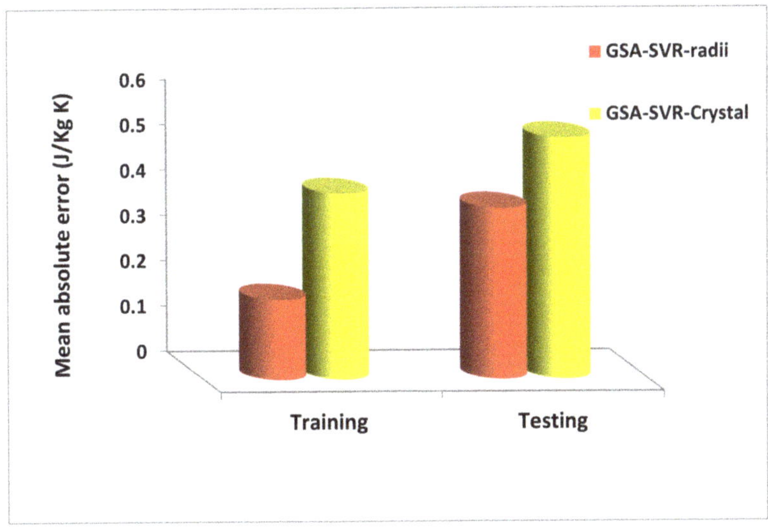

Figure 4. Comparison between the GSA-SVR-radii and GSA-SVR-crystal models on the basis of the mean absolute error.

Using MAE as the yardstick for measuring the generalization and future estimation capacity as presented in Figure 4, the developed GSA-SVR-radii model shows a better performance as compared to the GSA-SVR crystal model with a performance enhancement of 56.86% and 29.04% during the training and testing stages, respectively.

A similar improvement is presented in Figure 5 using RMSE as the performance measuring parameter. The prediction capacity of the developed GSA-SVR-radii model for the training dataset performs better than that of the GSA-SVR-crystal model with a performance improvement of 19.32%. while an improvement of 39.77% was obtained for the testing dataset. It should be noted that the testing stage of the developed GSA-SVR-radii model shows a better performance than the training phase of model development. This shows the robustness and uniqueness of the developed model in generalizing well to an unseen dataset. Another advantage of the developed GSA-SVR-radii model over the developed GSA-SVR-crystal model, aside from a better and excellent performance, is the ease of its descriptors. The descriptors for the developed GSA-SVR-radii model can be easily acquired without any prior experimental measurement, whereas XRD analysis must be performed before extracting the descriptors for the developed GSA-SVR-crystal model. Hence, the developed GSA-SVR-radii model allows pre-lab simulation and ultimately saves experimental time as well as other valuable resources.

Figure 6 presents the comparison between the ELM-GS-radii and ELM-GS-crystal models at the training and testing stages of model development using correlation coefficient as the measure of performance strength. The developed ELM-GS-crystal model performs better than the ELM-GS-radii model with a performance improvement of 5.29%, whereas, at the validation stage, the developed ELM-GS-radii model performs better than the ELM-GS-crystal model with a performance improvement of 5.07%. The comparison between the models is presented in Figure 7 with the mean absolute error as the performance measuring parameter. A similar trend of performance improvement is observed at the training and testing phases of model development.

The developed ELM-GS-crystal model shows an improvement of 14.08% as compared to the ELM-GS-radii model at the training stage, whereas the latter performs better than the former with a percentage improvement of 36.44% at the testing stage of model development. In Figure 8, which presents the comparison on the basis of RMSE, the ELM-GS-crystal model shows a percentage improvement of 19.11% over the developed ELM-GS-radii model at the training stage, whereas the

ELM-GS-radii model demonstrates a better performance of 39.35% at the testing stage. The actual values of each of the performance measuring parameters at different stages of model development are presented in Table 3. It is worth mentioning that although the developed ELM-GS-crystal model shows a better performance at the training phase, the excellent performance of the ELM-GS-radii model at the testing phase is more meritorious since the testing stage shows the future estimating capacity of the model.

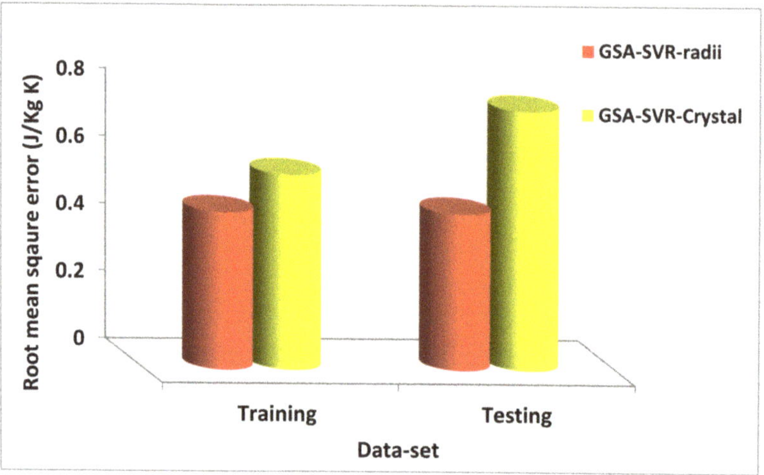

Figure 5. Comparison between the GSA-SVR-radii and GSA-SVR-crystal models on the basis of the root-mean-square error.

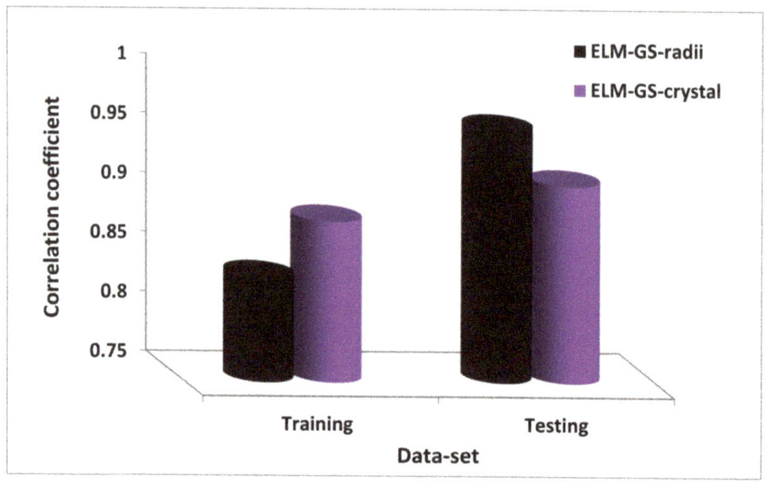

Figure 6. Comparison between the ELM-GS-radii and ELM-GS-crystal models on the basis of the correlation coefficient.

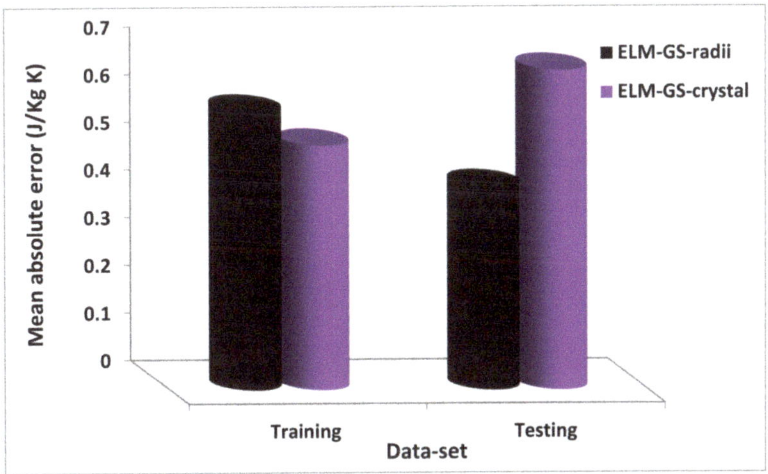

Figure 7. Comparison between the ELM-GS-radii and ELM-GS-crystal models on the basis of the mean absolute error.

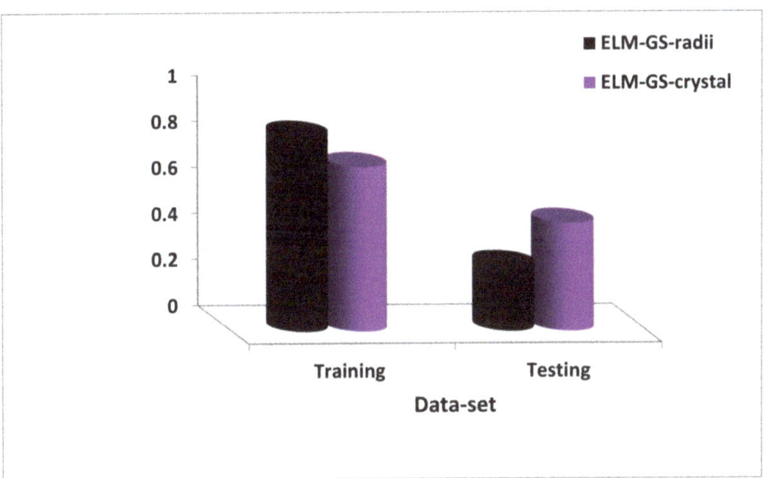

Figure 8. Comparison between the ELM-GS-radii and ELM-GS-crystal models on the basis of the root-mean-square error.

Table 3. Performance measuring parameters for the four developed models at different stages of model development.

Parameters	GSA-SVR-Radii		GSA-SVR-Crystal		ELM-GS-Radii		ELM-GS-Crystal	
	Training	Testing	Training	Testing	Training	Testing	Training	Testing
CC	0.9545	0.958	0.9317	0.9044	0.83777	0.961358	0.88462	0.914984
RMSE (J/Kg K)	0.4655	0.462	0.577	0.767	0.85233	0.285445	0.715575	0.46994
MAE (J/Kg K)	0.1766	0.376	0.4094	0.5299	0.586574	0.426137	0.51419	0.670479

The correlation cross-plots between the measured and predicted MMEC for training and testing sets of data are preesented in Figures 9 and 10, respectively. The values of the correlation coefficient for each of the developed models are also presented in the cross-plot for easy comparison. A direct impression about the precision and quality of the developed models can be easily inferred from the distribution and the alignment of the data points in the cross-plots.

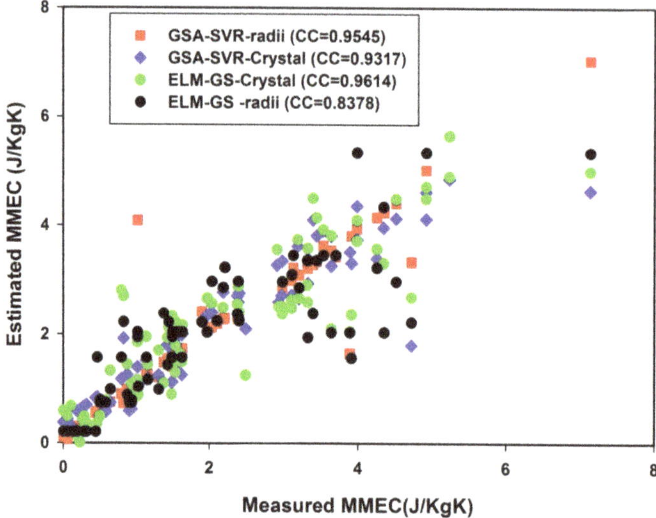

Figure 9. Correlation cross-plot between the measured and estimated MMEC for all developed models using the training set of data.

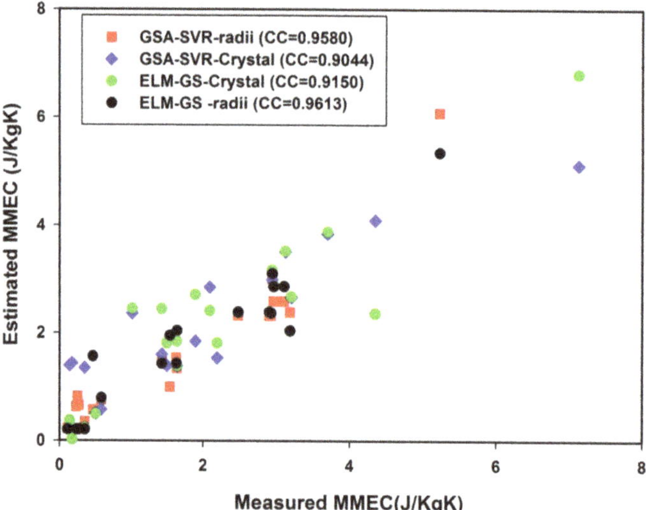

Figure 10. Correlation cross-plot between the measured and estimated MMEC for all developed models using the testing set of data.

4.3. Superiority of the Present Models as Compared to the Existing Models in the Literature

The generalization and predictive strength of the present models are compared with the existing models using mean absolute error and root-mean-square error metrics. The comparison between the performance of the present and existing models that employ the ionic radii and the concentrations of dopants as descriptors is presented in Figure 11 on the basis of the mean absolute error, while Figure 12 presents the same finding using the root-mean-square error as the performance measuring parameter.

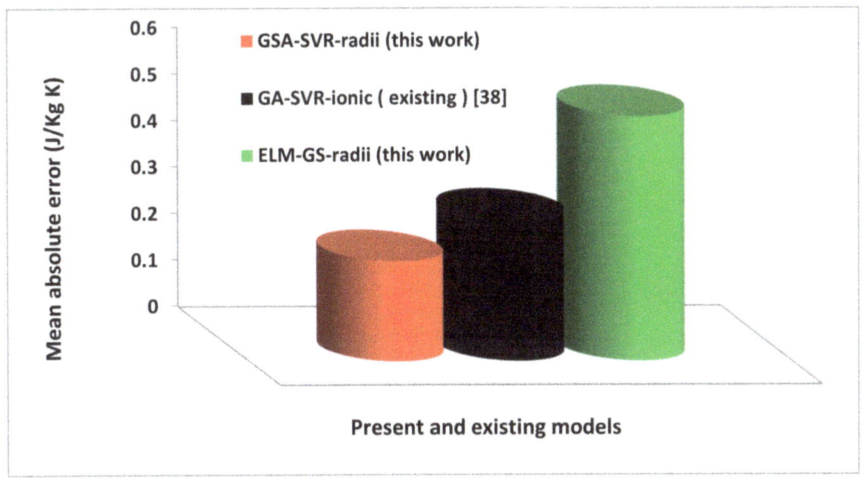

Figure 11. Comparison between the present and existing ionic radii-based models using the mean absolute error metric.

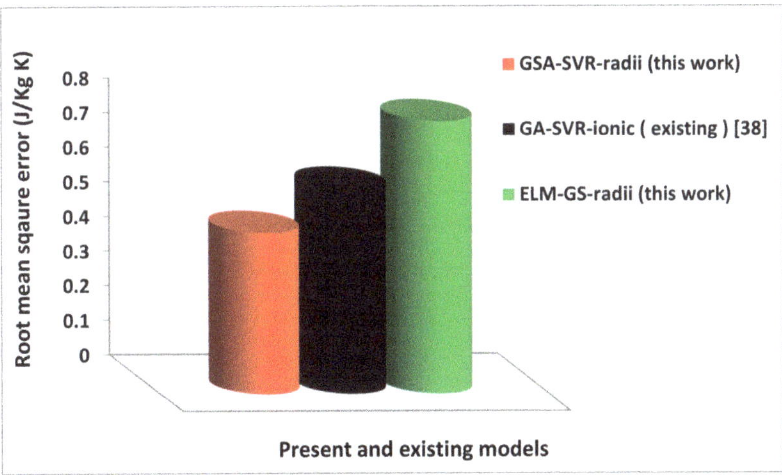

Figure 12. Comparison between the present and existing ionic radii-based models using the root-mean-square error metric.

The developed GSA-SVR-radii model performs better than the existing GA-SVR-ionic model [38] with a performance improvement of 28.85% as can be observed from Figure 11. The developed GSA-SVR-radii model also performs better than the developed ELM-GS-radii model with a performance of 58.86%. The superior performance demonstrated by the developed GSA-SVR-radii model can be attributed to the uniqueness of the Newtonian-based optimization algorithm in effectively selecting optimum hyperparameters of the SVR algorithm. Following a similar trend, the developed GSA-SVR-radii model performs better than the existing GA-SVR-ionic model [38] and the present ELM-GS-radii model with a performance improvement of 27.09% and 40.85%, respectively, on the basis of RMSE as presented in Figure 12.

Figures 13 and 14 present the comparison between the present and existing models that employ crystal lattice distortions as well as the applied magnetic field as the descriptors. The comparison presented in Figure 13 is based on the mean absolute error, whereas that of Figure 14 is based on the root-mean-square error. Table 4 also presents the actual values of each of the performance measuring parameters for the present and existing models.

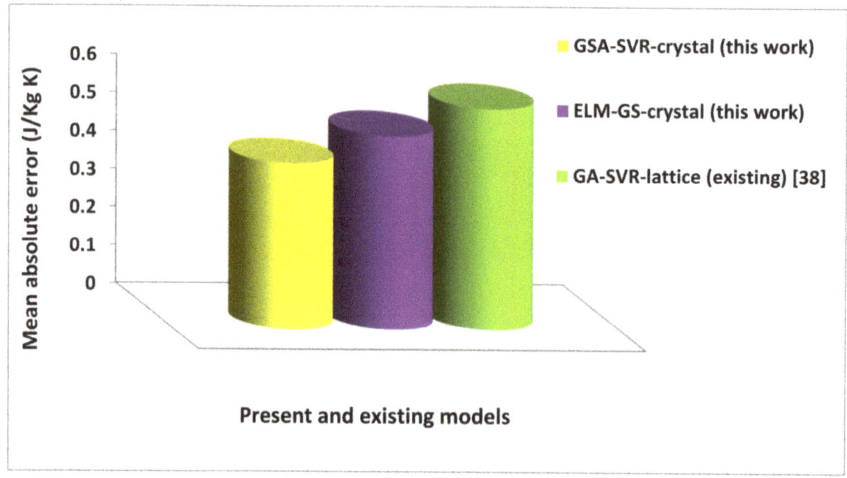

Figure 13. Comparison between the present and existing crystal lattice distortion-based models using the mean absolute error metric.

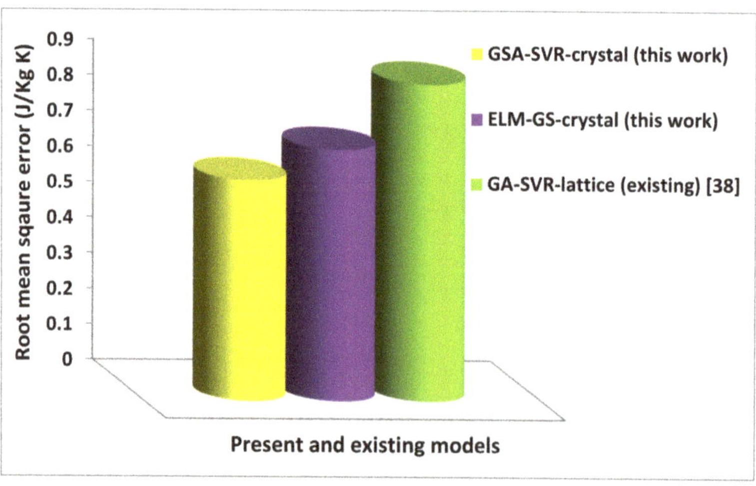

Figure 14. Comparison between the present and existing crystal lattice distortion-based models using the root-mean-square error metric.

Table 4. Performance comparison between the present and existing models.

	GA-SVR_Ionic (Existing) [38]	GA-SVR-Lattice (Existing) [38]	GSA-SVR-Radii (This Work)	GSA-SVR-Crystal (This Work)	ELM-GS-Radii (This Work)	ELM-GS-Crystal (This Work)
MAE (J/Kg K)	0.304311	0.578056	0.216517	0.433503	0.526348	0.50534
RMSE (J/Kg K)	0.590673	0.890676	0.464783	0.619666	0.785806	0.706786

The developed GSA-SVR-crystal model performs better than the existing GA-SVR-lattice model [38] with a performance improvement of 33.35% and 38.34% on the basis of MAE and RMSE, respectively. The developed ELM-GS-crystal model also performs better than the existing GA-SVR-ionic model with a performance enhancement of 14.39% and 20.65% using MAE and RMSE as performance measuring parameters, respectively. The observed precision of the developed crystal lattice distortion-based models over the existing model can be attributed to the strong mathematical background of the developed models as well as the excellent power of hybridization.

4.4. Investigating the Doping Effect of Fe on the Value of the MMEC of the $Pr_{0.6}Ca_{0.1}Sr_{0.3}Mn_{1-x}Fe_xO_3$ Manganite Compound Using the Developed GSA-SVR-Crystal Model

The influence of Fe dopants on the MMEC of the $Pr_{0.6}Ca_{0.1}Sr_{0.3}Mn_{1-x}Fe_xO_3$ manganite compound is presented in Table 5. It was observed that an increase in the concentration of Fe lowers the value of the MMEC. This observation as obtained using the developed GSA-SVR-crystal model conforms to the experimental observation [70]. The effect of the partial substitution of Fe in the Mn site in the $Pr_{0.6}Ca_{0.1}Sr_{0.3}Mn_{1-x}Fe_xO_3$ manganite compound lowers the value of the MMEC because the incorporation of Fe particles modifies the Mn/Fe-O distance and Mn-O-Mn angle, which subsequently leads to a weakening of the double-exchange interaction at the expense of the superexchange interaction [38].

Table 5. Comparison between the results of the developed GSA-SVR-crystal model and the measured values of the MMEC for different classes of manganite-based compounds.

Doped Manganite-Based Compounds	Measured MMEC (J/KgK)	Estimated MMEC Using GSA-SVR-Crystal (J/KgK)	Absolute Error
$La_{0.7}Ca_{0.25}Sr_{0.05}MnO_3$	1.65 [73]	1.5601	0.0899
$La_{0.67}Ca_{0.13}Ba_{0.2}Mn_{0.9}Co_{0.1}O_3$	1.93 [74]	1.5141	0.4159
$La_{0.6}Bi_{0.1}Sr_{0.3}Mn_{0.9}Cu_{0.1}O_3$	3.39 [75]	3.9562	0.5662
$La_{0.8}Na_{0.2}Mn_{0.94}Ni_{0.06}O_3$	3.70 [76]	4.1572	0.4572
$La_{0.065}Nd_{0.05}Ba_{0.3}Mn_{0.85}Cr_{0.15}O_3$	3.55 [77]	3.1356	0.4144
$Pr_{0.6}Ca_{0.1}Sr_{0.3}MnO_3$	3.64 [70]	3.8108	0.1708
$Pr_{0.6}Ca_{0.1}Sr_{0.3}Mn_{0.975}Fe_{0.025}O_3$	3.53 [70]	3.9058	0.3758
$Pr_{0.6}Ca_{0.1}Sr_{0.3}Mn_{0.95}Fe_{0.05}O_3$	3.7 [70]	3.845	0.145
$Pr_{0.6}Ca_{0.1}Sr_{0.3}Mn_{0.925}Fe_{0.075}O_3$	3.12 [70]	3.5018	0.3818

4.5. Implementation of the Developed Model for Determining the Maximum Magnetic Entropy Change of Different Classes of Doped Manganite

In order to further investigate the generalization and predictive strength of the developed model in determining the MMEC of manganite-based compounds, the developed GSA-SVR-crystal model was employed in determining the MMEC of several classes of manganite-based compounds and the obtained results were compared with the experimental values. The comparison is presented in Table 5. It is worth mentioning that the developed GSA-SVR-crystal model was only supplied with the descriptors, whereas the model implements the support vectors that it acquired during the training phase for its estimation. The results of the developed model agree well with the experimentally measured values. It has been reported that the experimental values of the MMEC suffer from an experimental uncertainty of approximately at least 5% [78] and that this may explain deviations of the results of the developed model from the measured values.

5. Conclusions and Recommendations

This work developed a grid search (GS)-based extreme learning machine (ELM) and hybrid gravitational search algorithm (GSA)-based support vector regression (SVR) for estimating the maximum magnetic entropy change (MMEC) of doped manganite-based compounds. The developed GSA-SVR-radii and ELM-GS-radii models employ ionic radii and concentrations of dopants at a specific value of an applied magnetic field as descriptors, whereas the developed GSA-SVR-crystal and ELM-GS-crystal models utilize the crystal lattice distortions due to the incorporation of dopants into the crystal lattice structure of the parent manganite compound as well as the value of the applied magnetic field as descriptors. The developed models perform better than the existing models in the literature using RMSE and MAE as performance measuring parameters. The better performance of the present models as compared to the existing models can be attributed to the sound mathematical background as well as the intrinsic ability of the models to precisely acquire all the intricacies, information and patterns linking the descriptors with the target. The precision of the developed models offers a green solution to the known pollutant-based refrigerants and opens up ways by which manganite-based compounds of the desired maximum magnetic entropy change can be predicted for possible laboratory fabrication and implementation.

Author Contributions: All authors have read and agreed to the published version of the manuscript. T.O.O. conceived, involved in data curation and wrote the original draft while S.M.I.S. designed the experiment, performed the experiment, analyzed the data, interpreted the data, involved in reviewing and editing. Both authors provided the materials for the experiment.

Funding: This research received no external funding.

Acknowledgments: The support provided by the University of Hafr Al Batin and Adekunle Ajasin University is acknowledged.

Conflicts of Interest: The authors declare no conflict of interest.

References

1. Hamdi, R.; Tozri, A.; Dhahri, E.; Bessais, L. Magnetocaloric properties and Landau theory of $Dy_{0.5}(Sr_{1-x}Ca_x)_{0.5}MnO_3$ (0≤ x≤0.3) manganites at cryogenic temperatures. *Chem. Phys. Lett.* **2017**, *680*, 94–100. [CrossRef]
2. Sellami-Jmal, E.; Ezaami, A.; Cheikhrouhou-Koubaa, W.; Cheikhrouhou, A. Investigation on physical properties in lanthanum vacancy of La0.65Ca0.35MnO3 elaborated at high temperature. *J. Magn. Magn. Mater.* **2018**, *465*, 762–767. [CrossRef]
3. Ayadi, F.; Ammar, S.; Nowak, S.; Cheikhrouhou-Koubaa, W.; Regaieg, Y.; Koubaa, M.; Monnier, J.; Sicard, L. Importance of the synthesis and sintering methods on the properties of manganite ceramics: The example of La 0.7 Ca 0.3 MnO 3. *J. Alloys Compd.* **2018**, *759*, 52–59. [CrossRef]
4. Das, K.; Banik, S.; Das, I. Large magnetocaloric effect in geometrically frustrated polycrystalline ErMnO 3 compound at cryogenic temperature. *Phys. B Condens. Matter* **2018**, *533*, 46–49. [CrossRef]
5. Guorui, X..; Wei, H.; Peiqi, C.; Xiaowei, W. Effect of Nd-substitution on the structural, magnetic and magnetocaloric properties of La0.67-xNdxCa0.13Ba0.2MnO3 manganites. *J. Magn. Magn. Mater.* **2019**, *491*, 133–142.
6. Owolabi, T.O.; Akande, K.O.; Olatunji, S.O.; Aldhafferi, N.; Alqahtani, A. Support Vector Regression Ensemble for Effective Modeling of Magnetic Ordering Temperature of Doped Manganite in Magnetic Refrigeration. *J. Low Temp. Phys.* **2019**, *195*, 179–201. [CrossRef]
7. Linh, D.C.; Ha, N.T.; Duc, N.H.; Nam, L.H.G.; Bau, L.V.; An, N.M.; Yu, S.-C.; Tran, D.T. Na-doped La 0.7 Ca 0.3 MnO 3 compounds exhibiting a large magnetocaloric effect near room temperature. *Phys. B Condens. Matter* **2018**, *532*, 155–160. [CrossRef]
8. Das, S.; Roychowdhury, P.; De, S.; Roy, A.; Chatterjee, S.; De, K. Magnetic and electrical transport of the cation-deficient LaMnO 3: Common origin for both Sr-doping and self-doping effects. *Phys. B Condens. Matter* **2018**, *544*, 17–22. [CrossRef]
9. Yen, P.D.H.; Dung, N.T.; Tran, D.T.; Yu, S.-C. Magnetic properties and magnetocaloric effect of Sr-doped Pr0.7Ca0.3MnO3 compounds. *Curr. Appl. Phys.* **2018**, *18*, 1280–1288. [CrossRef]
10. Aziz, F.; Chandra, M.; Das, S.; Prajapat, M.; Mavani, K. Structural and metamagnetic transitions in thin films of Ce-doped Pr0.5Ca0.5MnO3 manganites. *Thin Solid Films* **2016**, *615*, 338–344. [CrossRef]
11. Sudakshina, B.; Arun, B.; Chandrasekhar, K.D.; Yang, H.D.; Vasundhara, M. Structural and magnetic properties of Nd 0.67 Ba 0.33 MnO 3 manganites with partial replacement of Fe and Cu at Mn-site. *Phys. B Condens. Matter* **2018**, *539*, 14–20. [CrossRef]

12. Asmira, N.; Ibrahim, N.; Mohamed, Z.; Yahya, A. Effect of Cr 3+ substitution at Mn-site on electrical and magnetic properties of charge ordered Bi 0.3 Pr 0.3 Ca 0.4 MnO 3 manganites. *Phys. B Condens. Matter* **2018**, *544*, 34–46. [CrossRef]
13. Kataria, B.; Solanki, P.; Pandya, D.; Solanki, P.; Shah, N. Role of antimony in the charge transport mechanisms for La 0.67 Ca 0.33 Mn 1−x Sb x O 3 manganites. *Phys. B Condens. Matter* **2018**, *541*, 43–49. [CrossRef]
14. Poojary, T.; Babu, P.; Sanil, T.; Daivajna, M.D. Effect of gadolinium dopant on structural, magneto-transport, magnetic and thermo-power of Pr 0.8 Sr 0.2 MnO 3. *Solid State Commun.* **2018**, *275*, 35–42. [CrossRef]
15. Makni-Chakroun, J.; M'Nassri, R.; Cheikhrouhou-Koubaa, W.; Koubaa, M.; Chniba-Boudjada, N.; Cheikhrouhou, A. Effect of A-site deficiency on investigation of structural, magnetic and magnetocaloric behaviors for (LaSr)-lacunar manganites. *Chem. Phys. Lett.* **2018**, *707*, 61–70. [CrossRef]
16. Akça, G.; Çetin, S.K.; Güneş, M.; Ekicibil, A. Magnetocaloric properties of (La1−xPrx)0.85K0.15MnO3 (x=0.0, 0.1, 0.3 and 0.5) perovskite manganites. *Ceram. Int.* **2016**, *3*, 3–10. [CrossRef]
17. Selmi, A.; M'Nassri, R.; Cheikhrouhou-Koubaa, W.; Boudjada, N.C.; Cheikhrouhou, A. Effects of partial Mn-substitution on magnetic and magnetocaloric properties in Pr0.7Ca0.3Mn0.95X0.05O3 (Cr, Ni, Co and Fe) manganites. *J. Alloys Compd.* **2015**, *619*, 627–633. [CrossRef]
18. Vadnala, S.; Asthana, S. Magnetocaloric effect and critical field analysis in Eu substituted La0.7-xEuxSr0.3MnO3(x = 0.0, 0.1, 0.2, 0.3) manganites. *J. Magn. Magn. Mater.* **2018**, *446*, 68–79. [CrossRef]
19. Bocarsly, J.D.; Levin, E.E.; Garcia, C.A.C.; Schwennicke, K.; Wilson, S.D.; Seshadri, R. A Simple Computational Proxy for Screening Magnetocaloric Compounds. *Chem. Mater.* **2017**, *29*, 1613–1622. [CrossRef]
20. Zarkevich, N.; Johnson, D.D. Reliable thermodynamic estimators for screening caloric materials. *J. Alloys Compd.* **2019**, *802*, 712–722. [CrossRef]
21. Owolabi, T.; Akande, K.O.; Olatunji, S.O.; Alqahtani, A.; Aldhafferi, N. Estimation of Curie temperature of manganite-based materials for magnetic refrigeration application using hybrid gravitational based support vector regression. *AIP Adv.* **2016**, *6*, 105009. [CrossRef]
22. Owolabi, T.; Oloore, L.E.; Akande, K.O.; Olatunji, S.O. Modeling of magnetic cooling power of manganite-based materials using computational intelligence approach. *Neural Comput. Appl.* **2017**, *31*, 1291–1298. [CrossRef]
23. Owolabi, T.; Akande, K.O.; Olatunji, S.O.; Alqahtani, A.; Aldhafferi, N. Incorporation of GSA in SBLLM-based neural network for enhanced estimation of magnetic ordering temperature of manganite. *J. Intell. Fuzzy Syst.* **2017**, *33*, 1225–1233. [CrossRef]
24. Owolabi, T.; Akande, K.O.; Olatunji, S.O.; Alqahtani, A.; Aldhafferid, N. Modeling of Curie temperature of manganite for magnetic refrigeration application using manual search and hybrid gravitational-based support vector regression. *Soft Comput.* **2017**, *22*, 3023–3032. [CrossRef]
25. Owolabi, T.; Akande, K.O.; Olatunji, S.O.; Aldhafferi, N.; Alqahtani, A. Ensemble-Based Support Vector Regression with Gravitational Search Algorithm Optimization for Estimating Magnetic Relative Cooling Power of Manganite Refrigerant in Magnetic Refrigeration Application. *J. Supercond. Nov. Magn.* **2018**, *32*, 2107–2118. [CrossRef]
26. Vapnik, V.N. *Statistical Learning Theory*; Wiley-Interscience: New York, NY, USA, 1998.
27. Oloore, L.; Owolabi, T.; Fayose, S.; Adegoke, M.; Akande, K.; Olatunji, S. Modeling of semiconductors refractive indices using hybrid chemometric model. *Model. Meas. Control. A* **2018**, *91*, 95–103. [CrossRef]
28. Owolabi, T.; Akande, K.O.; Olatunji, S.O. Estimation of average surface energies of transition metal nitrides using computational intelligence technique. *Soft Comput.* **2016**, *21*, 6175–6182. [CrossRef]
29. Zhang, H.-R.; Zhang, Y.; Dai, D.-B.; Cao, M.; Shen, W.-F. Modelling and optimization of the superconducting transition temperature. *Mater. Des.* **2016**, *92*, 371–377. [CrossRef]
30. Cai, C.; Xiao, T.; Tang, J.; Huang, S. Analysis of process parameters in the laser deposition of YBa2Cu3O7 superconducting films by using SVR. *Phys. C Supercond.* **2013**, *493*, 100–103. [CrossRef]
31. Motamedi, S.; Shamshirband, S.; Hashim, R.; Petković, D.; Roy, C. RETRACTED: Estimating unconfined compressive strength of cockle shell–cement–sand mixtures using soft computing methodologies. *Eng. Struct.* **2015**, *98*, 49–58. [CrossRef]
32. Behnood, A.; Verian, K.P.; Modiri-Gharehveran, M. Evaluation of the splitting tensile strength in plain and steel fiber-reinforced concrete based on the compressive strength. *Constr. Build. Mater.* **2015**, *98*, 519–529. [CrossRef]

33. Rashedi, E.; Nezamabadi-Pour, H.; Saryazdi, S. GSA: A Gravitational Search Algorithm. *Inf. Sci.* **2009**, *179*, 2232–2248. [CrossRef]
34. Huang, G.-B.; Zhu, Q.-Y.; Siew, C.-K. Extreme learning machine: Theory and applications. *Neurocomputing* **2006**, *70*, 489–501. [CrossRef]
35. Yan, C.; Qi, J.; Ma, J.; Tang, H.; Zhang, T.; Li, H. Determination of carbon and sulfur content in coal by laser induced breakdown spectroscopy combined with kernel-based extreme learning machine. *Chemom. Intell. Lab. Syst.* **2017**, *167*, 226–231. [CrossRef]
36. Owolabi, T.; Gondal, M. Development of hybrid extreme learning machine based chemo-metrics for precise quantitative analysis of LIBS spectra using internal reference pre-processing method. *Anal. Chim. Acta* **2018**, *1030*, 33–41. [CrossRef]
37. Wang, Y.G.; Cao, F.; Yuan, Y. A study on effectiveness of extreme learning machine. *Neurocomputing* **2011**, *74*, 2483–2490. [CrossRef]
38. Owolabi, T. Modeling the magnetocaloric effect of manganite using hybrid genetic and support vector regression algorithms. *Phys. Lett. A* **2019**, *383*, 1782–1790. [CrossRef]
39. Boser, B.E.; Guyon, I.M.; Vapnik, V.N. A training algorithm for optimal margin classifiers. In Proceedings of the 5th Annual ACM Workshop on Computational Learning Theory, Pittsburgh, PA, USA, 27–29 July 1992; pp. 144–152.
40. Basak, D.; Pal, S.; Patranabis, D.C. Support Vector Regression. *Neural Inf. Process. Lett. Rev.* **2007**, *11*, 203–224.
41. Peng, J.; Li, L. Support vector regression in sum space for multivariate calibration. *Chemom. Intell. Lab. Syst.* **2014**, *130*, 14–19. [CrossRef]
42. Aldhafferi, N.; Owolabi, T.O.; Akande, K.O.; Olatunji, S.O.; Alqahtani, A. Support vector regression approach of modeling the superconducting transition temperature of borocarbide-based superconductors. *ARPN J. Eng. Appl. Sci.* **2018**, *13*, 1900–1905.
43. Owolabi, T.; Akande, K.O.; Olatunji, S.O. Estimation of surface energies of hexagonal close packed metals using computational intelligence technique. *Appl. Soft Comput.* **2015**, *31*, 360–368. [CrossRef]
44. Owolabi, T.; Suleiman, M.A.; Adeyemo, H.; Akande, K.O.; Alhiyafi, J.; Olatunji, S.O. Estimation of minimum ignition energy of explosive chemicals using gravitational search algorithm based support vector regression. *J. Loss Prev. Process. Ind.* **2019**, *57*, 156–163. [CrossRef]
45. Adeyemo, H.B.; Owolabi, T.O.; Suleiman, M.A.; Akande, A.J.; Fayose, S.; Olatuji, S.O. *Heliyon Hybrid chemometric approach for estimating the heat of detonation of aromatic energetic compounds. Heliyon* **2019**, *5*, 1–8.
46. Owolabi, T. Determination of the Velocity of Detonation of Primary Explosives Using Genetically Optimized Support Vector Regression. *Propellants Explos. Pyrotech.* **2019**, 1–12. [CrossRef]
47. Yildiz, B.S.; Lekesiz, H.; Yildiz, A.R.; Yıldız, B.S. Structural design of vehicle components using gravitational search and charged system search algorithms. *Mater. Test.* **2016**, *58*, 79–81. [CrossRef]
48. Sabri, N.M.; Puteh, M.; Mahmood, M.R. A review of gravitational search algorithm. *Int. J. Adv. Soft Comput. Appl.* **2013**, *5*.
49. Owolabi, T.; Gondal, M.A. A hybrid intelligent scheme for estimating band gap of doped titanium dioxide semiconductor using crystal lattice distortion. *Comput. Mater. Sci.* **2017**, *137*, 249–256. [CrossRef]
50. Huang, G.-B.; Chen, L. Convex incremental extreme learning machine. *Neurocomputing* **2007**, *70*, 3056–3062. [CrossRef]
51. Owolabi, T.; Gondal, M.A. Quantitative analysis of LIBS spectra using hybrid chemometric models through fusion of extreme learning machines and support vector regression. *J. Intell. Fuzzy Syst.* **2018**, 1–10. [CrossRef]
52. ElGhoul, A.; Krichene, A.; Boudjada, N.C.; Boujelben, W. Rare earth effect on structural, magnetic and magnetocaloric properties of La0.75Ln0.05Sr0.2MnO3 manganites. *Ceram. Int.* **2018**, *44*, 12723–12730. [CrossRef]
53. Koubaa, M.; Regaieg, Y.; Koubaa, W.C.; Cheikhrouhou, A.; Ammar-Merah, S.; Herbst, F. Magnetic and magnetocaloric properties of lanthanum manganites with monovalent elements doping at A-site. *J. Magn. Magn. Mater.* **2011**, *323*, 252–257. [CrossRef]
54. Nedelko, N.; Lewińska, S.; Pashchenko, A.; Radelytskyi, I.; Diduszko, R.; Zubov, É.E.; Lisowski, W.; Sobczak, J.; Dyakonov, K.; Ślawska-Waniewska, A.; et al. Magnetic properties and magnetocaloric effect in La0.7Sr0.3−xBixMnO3 manganites. *J. Alloys Compd.* **2015**, *640*, 433–439. [CrossRef]

55. Arun, B.; Athira, M.; Akshay, V.R.; Sudakshina, B.; Mutta, G.R.; Vasundhara, M. Investigation on the structural, magnetic and magnetocaloric properties of nanocrystalline Pr-deficient Pr1−xSrxMnO3−δmanganites. *J. Magn. Magn. Mater.* **2018**, *448*, 322–331. [CrossRef]

56. Saleh, J.A.; Sarsari, I.A.; Kameli, P.; Salamati, H. Influence of Al-doping on the structural, magnetic, and electrical properties of La0.8Ba0.2Mn1−xAlxO3(0 ≤ x ≤ 0.25) manganites. *J. Magn. Magn. Mater.* **2018**, *465*, 339–347. [CrossRef]

57. Choudhary, Y.; Mangavati, S.; Patil, S.; Rao, A.; Nagaraja, B.; Thomas, R.; Okram, G.; Kini, S.G. Effect of rare-earth substitution at La-site on structural, electrical and thermoelectric properties of La0.7−xRExSr0.3MnO3 compounds (x = 0, 0.2, 0.3; RE = Eu, Gd, Y). *J. Magn. Magn. Mater.* **2018**, *451*, 110–120. [CrossRef]

58. Liu, Z.; Lin, W.; Zhou, K.; Yan, J. Effect of Cu doping on the structural, magnetic and magnetocaloric properties of La0.7Sr0.25Na0.05Mn1−xCuxO3 manganites. *Ceram. Int.* **2018**, *44*, 2797–2802. [CrossRef]

59. Ben Khlifa, H.; Othmani, S.; Chaaba, I.; Tarhouni, S.; Cheikhrouhou-Koubaa, W.; Koubaa, M.; Cheikhrouhou, A.; Hlil, E. Effect of K-doping on the structural, magnetic and magnetocaloric properties of Pr0.8Na0.2–K MnO3 (0 ≤ x ≤ 0.15) manganites. *J. Alloys Compd.* **2016**, *680*, 388–396. [CrossRef]

60. Iqbal, M.; Khan, M.N.; Khan, A.A. Structural, magnetic, magnetocaloric and critical behavior studies in the vicinity of the paramagnetic to ferromagnetic phase transition temperature in LaMnO3+δcompound. *J. Magn. Magn. Mater.* **2018**, *465*, 670–677. [CrossRef]

61. Snini, K.; Ben Jemaa, F.; Mohamed, E.; Hlil, E. Structural, magnetic and magnetocaloric investigations in Pr0.67Ba0.22Sr0.11Mn1-xFexO3 (0 ≤ x ≤ 0.15) manganite oxide. *J. Alloys Compd.* **2018**, *739*, 948–954. [CrossRef]

62. Tejaswini, B.P.; Daivajna, M.D. Thrupthi Structural, electrical, magnetic and thermal properties of Pr0.8-xDyxSr0.2MnO3 with (x = 0, 0.2 and 0.25). *J. Alloys Compd.* **2018**, *741*, 97–105. [CrossRef]

63. Hira, U.; Sher, F. Structural, magnetic and high-temperature thermoelectric properties of La0.4Bi0.4Ca0.2Mn1−Co O3 (0 ≤ x ≤ 0.3) perovskites. *J. Magn. Magn. Mater.* **2018**, *452*, 64–72. [CrossRef]

64. Elyana, E.; Mohamed, Z.; Kamil, S.; Supardan, S.; Chen, S.; Yahya, A. Revival of ferromagnetic behavior in charge-ordered Pr0.75Na0.25MnO3 manganite by ruthenium doping at Mn site and its MR effect. *J. Solid State Chem.* **2018**, *258*, 191–200. [CrossRef]

65. Ghodhbane, S.; Tka, E.; Dhahri, J.; Hlil, E. A large magnetic entropy change near room temperature in La0.8Ba0.1Ca0.1Mn0.97Fe0.03O3 perovskite. *J. Alloys Compd.* **2014**, *600*, 172–177. [CrossRef]

66. Selmi, A.; M'Nassri, R.; Cheikhrouhou-Koubaa, W.; Boudjada, N.C.; Cheikhrouhou, A. Influence of transition metal doping (Fe, Co, Ni and Cr) on magnetic and magnetocaloric properties of Pr0.7Ca0.3MnO3 manganites. *Ceram. Int.* **2015**, *41*, 10177–10184. [CrossRef]

67. Wang, Z.; Jiang, J. Magnetic entropy change in perovskite manganites La0.7A0.3MnO3 La0.7A0.3Mn0.9Cr0.1O3 (A = Sr, Ba, Pb) and Banerjee criteria on phase transition. *Solid State Sci.* **2013**, *18*, 36–41. [CrossRef]

68. Selmi, A.; M'Nassri, R.; Cheikhrouhou-Koubaa, W.; Boudjada, N.C.; Cheikhrouhou, A. The effect of Co doping on the magnetic and magnetocaloric properties of Pr0.7Ca0.3Mn1−xCoxO3 manganites. *Ceram. Int.* **2015**, *41*, 7723–7728. [CrossRef]

69. Oumezzine, E.; Hcini, S.; Hlil, E.-K.; Dhahri, E.; Oumezzine, M. Effect of Ni-doping on structural, magnetic and magnetocaloric properties of La 0.6 Pr 0.1 Ba 0.3 Mn 1− x Ni x O 3 nanocrystalline manganites synthesized by Pechini sol–gel method. *J. Alloys Compd.* **2014**, *615*, 553–560. [CrossRef]

70. Mahjoub, S.; Baazaoui, M.; M'Nassri, R.; Rahmouni, H.; Boudjada, N.C.; Oumezzine, M. Effect of iron substitution on the structural, magnetic and magnetocaloric properties of Pr0.6Ca0.1Sr0.3Mn1−xFexO3 (0 ≤ x ≤ 0.075) manganites. *J. Alloys Compd.* **2014**, *608*, 191–196. [CrossRef]

71. Mleiki, A.; Othmani, S.; Cheikhrouhou-Koubaa, W.; Koubaa, A.; Cheikhrouhou, A.; Hlil, E.K. Effect of praseodymium doping on the structural, magnetic and magnetocaloric properties of Sm0.55Sr0.45MnO3 manganite. *J. Alloys Compd.* **2015**, *645*, 559–565. [CrossRef]

72. Kossi, S.E.; Ghodhbane, S.; Dhahri, J.; Hlil, E.K. The impact of disorder on magnetocaloric properties in Ti-doped manganites of La0.7Sr0.25Na0.05Mn(1-x)TixO3 (0≤x ≤0.2). *J. Magn. Magn. Mater.* **2015**, *395*, 134–142. [CrossRef]

73. Taboada-Moreno, C.; Jesús, F.S.-D.; Pedro-García, F.; Cortés-Escobedo, C.; Betancourt-Cantera, J.; Ramírez-Cardona, M.; Bolarín-Miró, A. Large magnetocaloric effect near to room temperature in Sr doped La0.7Ca0.3MnO3. *J. Magn. Magn. Mater.* **2020**, *496*, 165887. [CrossRef]
74. Xiao, G.; He, W.; Yang, T.; Huang, G.; Wang, T.; Huang, J. Effect of Co-doping on structural, magnetic and magnetocaloric properties of La 0.67 Ca 0.13 Ba 0.2 Mn 1-x Co x O 3 (x = 0, 0.02, 0.04, 0.06, 0.08, 0.1) manganites. *Curr. Appl. Phys.* **2019**, *19*, 424–435. [CrossRef]
75. Bouzaiene, E.; Dhahri, A.; Dhahri, J.; Hlil, E.; Bajahzar, A. Effect of A-site-substitution on structural, magnetic and magnetocaloric properties in La0.7Sr0.3 Mn0.9Cu0.1O3 manganite. *J. Magn. Magn. Mater.* **2019**, *491*, 165540. [CrossRef]
76. Hassayoun, O.; Baazaoui, M.; Laouyenne, M.; Hosni, F.; Hlil, E.; Oumezzine, M.; Farah, K. Magnetocaloric effect and electron paramagnetic resonance studies of the transition from ferromagnetic to paramagnetic in La0.8Na0.2Mn1-xNixO3 (0≤x≤0.06). *J. Phys. Chem. Solids* **2019**, *135*, 109058. [CrossRef]
77. Mahjoub, S.; M'Nassri, R.; Baazaoui, M.; Hlil, E.; Oumezzine, M. Tuning magnetic and magnetocaloric properties around room temperature via chromium substitution in La0.65Nd0.05Ba0.3MnO3 system. *J. Magn. Magn. Mater.* **2019**, *481*, 29–38. [CrossRef]
78. Pecharsky, V.K.; Gschneidner, K.A. Magnetocaloric effect from indirect measurements: Magnetization and heat capacity. *J. Appl. Phys.* **1999**, *565*. [CrossRef]

Article

Reactively Synthesized Porous Ti_3SiC_2 Compound and Its Mechanical Properties with Different Apertures

Yao Jiang [1,*], Xinli Liu [2], Haiyan Gao [1] and Yuehui He [1]

[1] State Key Laboratory of Powder Metallurgy, Central South University, Changsha 410083, China; gaohy@csu.edu.cn (H.G.); yuehui@csu.edu.cn (Y.H.)
[2] School of Materials Science and Engineering, Central South University, Changsha 410083, China; liuxinli@csu.edu.cn
* Correspondence: jiangyao@csu.edu.cn

Received: 14 January 2020; Accepted: 1 February 2020; Published: 2 February 2020

Abstract: Reactively synthesized porous Ti_3SiC_2 with different pore sizes was prepared using TiH_2, Si and graphite powders as starting materials. The effect of pore size on the flexural stress–strain relationship, bending strength and flexural elastic modulus were investigated. The results show that the synthesized porous Ti_3SiC_2 intermetallic compounds have a characteristic of a high-purity MAX phase with typical laminate microstructure. When the average pore size decreases from 21.8 to 2.1 μm, the volume content of Ti_3SiC_2 phase ranges from 96.9% to 99.6%, and the porosity is in the range of 49.9% to 54.1%. The flexural stress–strain curves of porous Ti_3SiC_2 show a characteristic of two stages of elastic deformation and fracture. The flexural modulus is in the range of 13 to 70 GPa, which increases rapidly with further decrease of the pore size. A relation similar to the Hall–Petch equation between the mechanical property and the pore size was investigated for the porous material.

Keywords: Ti_3SiC_2; intermetallic compound; porous material; mechanical property; pore size; elastic modulus

1. Introduction

Porous Ti_3SiC_2 [1–5], as a new type of porous intermetallic compound with MAX phase, has the performance advantages of porous metals and ceramics, such as excellent environmental corrosion resistivity [1], high temperature stability [2] as well as good thermal shock resistance and machining performance [3]. In addition, porous Ti_3SiC_2 prepared by reactive synthesis of elemental powders [6–8] also shows a good open-pore structure and pore-structure stability [1,2].

Compared with dense material, the mechanical strength of porous material decreases significantly [9–11]. The mechanical model [8,12], microstructure [8] and the effect of porosity on mechanical properties [12,13] of porous Ti_3SiC_2 have been investigated, which is classified as a kinking nonlinear elastic solid [14,15]. The strength and stiffness of porous Ti_3SiC_2 decrease with increasing the porosity [12]. In the case of porous MAX phases prepared with space holder with the size of several hundred microns [4,5,13], it was found that the porosity and the pore type were the main factors influencing the elastic properties [13]. On the other hand, porous Ti_3SiC_2, due to its unique pore structure and material properties, can be well applied in biological separation [16], chemical filtration and hydrometallurgy [1], etc., where the mechanical properties of porous materials should meet the requirements of the working conditions [3]. The pore size is an important pore structure parameter determining the filtration accuracy, and its influence on mechanical properties is an important design basis for porous materials in the filtration application. To the best of our knowledge, investigations on the effect of pore size of porous Ti_3SiC_2 on its strength and stiffness have not been reported so far.

Currently, the relationship between porosity and strength of porous materials can be simulated by the Barsen equation [17]. The Hall–Petch equation can be used to reveal the relation between the strength or hardness and the grain size of single phase material [18–20] or composite material [21–23]. In this work, a series of porous Ti_3SiC_2 compounds with different pore sizes and approximately the same porosities and phase purities were prepared by the reactive synthesis of elemental powders through a pressure less sintering process. The Hall–Petch relationship between the pore size and the bending strength was studied, and the effect of pore size on the flexural modulus of porous Ti_3SiC_2 was discussed.

2. Experimental Procedure

Porous Ti_3SiC_2 compounds were prepared using commercially available TiH_2, Si and graphite powders as the starting materials by a pressure less sintering process [1]. TiH_2 powder can be prepared by hydrogenation of titanium sponge and subsequent mechanical pulverization process [24]. The median diameter of TiH_2 powder is in the range of 10 to 180 μm, which was utilized to fabricate porous material with different pore sizes. The median diameters of Si and graphite powders are in the ranges of 10–15 μm and 5–10 μm, respectively. The purities of all the used powders were above 99.0 wt.%, and were mixed according to the atomic ratio of Ti, Si, and C of 3:1.2:2. An excess of 20% Si was added to compensate for its high temperature volatilization loss [1,3]. Analytically pure stearic acid was used as forming agent and lubricant. Cold forming procedure was carried out by means of uniaxial die pressing with the pressure of 100 MPa. The prepared compacts are cuboids with dimensions of length, width and thickness of 30 mm, 6–7 mm and 2–3 mm, respectively. A multi-stage heating process under vacuum was utilized for the sintering of compacts [3]. The sintering procedure diagram is shown in Figure 1. The low-temperature sintering process at 350 °C is designed to decompose and remove the organic additives from the compacts, which is followed by a heating stage at 650 °C to dehydrogenate TiH_2. At the intermediate sintering stage at 1000 °C, the binary intermediate phases including TiC and Ti_5Si_3 are gradually formed through the solid phase diffusion [3,6], which is conducive to the subsequent synthesis of Ti_3SiC_2 [3]. The final sintering temperature is 1350 °C for 3 h with the vacuum pressure of 10^{-1}–10^{-3} Pa. The heating rate is controlled at 5–10 °C/min. After the pressure less sintering, the compacts are cooled with the furnace.

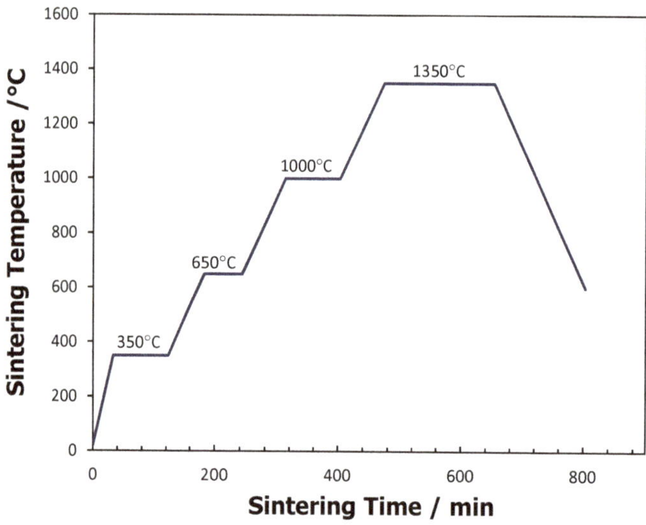

Figure 1. Temperature-time diagram for the sintering procedure of cold pressed compacts.

The phase compositions of the compacts after sintering at 1350 °C for 3 h were analyzed by X-ray diffraction (XRD: Dmax 2500VB, Tokyo, Japan) with a Cu Kα source at room temperature. The corresponding XRD data processing and analysis was carried out by Jade5 software (MDI, California, USA). The mechanical properties and the relation between flexural stress and strain of porous Ti_3SiC_2 were measured according to the three point flexural test by an electric servo-hydraulic material test system (Instron3369, Instron Ltd., High Wycombe, UK). The sintered compacts with the rectangular section mentioned above were used as the samples for the bending strength test, in which the span and the displacement rate are 25 mm and 0.02 mm/min, respectively. The mechanical strengths are the average of the three replicates used. The pore size was determined by the bubble point method (FBP-III) [6]. The open and overall porosities were calculated through the pycnometry method.

3. Results and Discussion

The XRD patterns of porous Ti_3SiC_2 with three different average pore sizes are shown in Figure 2. The phases of these porous materials are composed of Ti_3SiC_2 phase (a = 3.062 Å and c = 17.637 Å, Space Group P63/mmc, PDF 40-1132) and a small amount of TiC impurity phase (a = 4.327 Å, Space Group Fm-3m, PDF 65-0242). According to the standard additive method [25], the Ti_3SiC_2 phase volume contents of the porous compounds with average pore sizes of 2.1, 12.9 and 21.8 μm have been calculated to be 99.6%, 98.8% and 96.9%, respectively, showing a high-purity characteristic. The MAX phase content of porous Ti_3SiC_2 increases to a certain extent as the pore size reduces. The phase composition of porous Ti_3SiC_2 compounds prepared by reactive synthesis has been investigated to be influenced by the raw material type [1], the synthesis parameters [3], the material composition and alloying elements [7] and the material granularity [16]. The addition of TiH_2 was proved to improve the phase purity of the synthesized Ti_3SiC_2 MAX phase, which was attributed to the low oxygen content and high reactivity of elemental Ti generated from the dehydrogenation of TiH_2 [1,3].

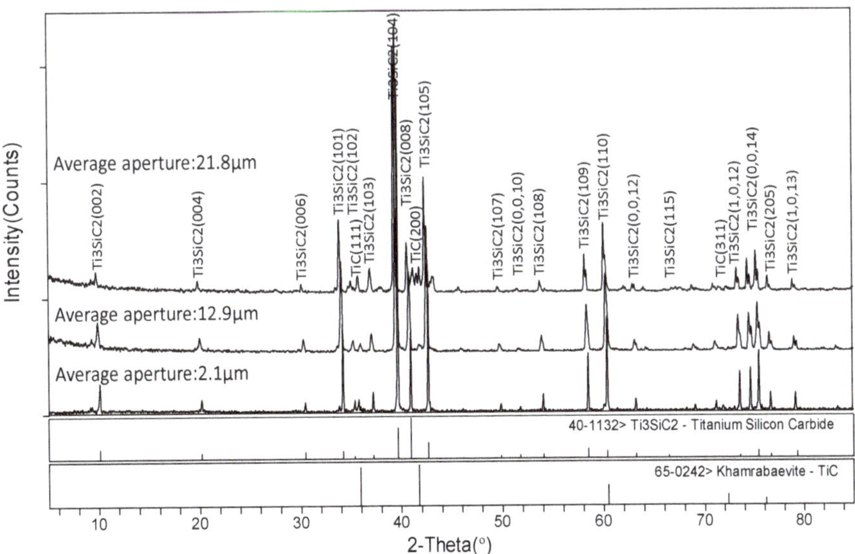

Figure 2. X-ray diffraction (XRD) patterns of porous Ti_3SiC_2 with different average pore sizes.

In the preparation process of porous Ti_3SiC_2 by the reactive synthesis of TiH_2, Si and graphite, the pore size of the resultant porous material has a strict linear relation with the powder particle size of TiH_2, and the porosity is kept essentially constant when the pore size changes [6]. Therefore, the pore size of porous Ti_3SiC_2 can be adjusted by changing the powder size of TiH_2. In fact, the synthesized porous Ti_3SiC_2 in this work has the average pore size in the range of 2.1–21.8 μm with the open and overall porosities in the ranges of 44.6%–48.5% and 49.9%–54.1%, respectively. The decrease of pore size means the increase of specific surface area of the reactant under the condition that porosity is basically constant, which helps to enhance the surface diffusion during the synthesis process.

Figure 3 reveals the pore structure morphology and the lamellar microstructure of porous Ti_3SiC_2. The synthesized Ti_3SiC_2 compound has the characteristics of rich porous structure and high specific surface area, as shown in Figure 3a and b with the average apertures of 6.2 μm and 3.0 μm, respectively. A large number of interconnected pores are generated in the sintered compact with the aperture at the micron level. A few coarse pores exist locally, which is related to the original particle clearance in the compacts [7]. The prepared Ti_3SiC_2 phase skeleton generally has a smooth surface, which is conducive to the permeability of the porous material. Further observations reveal that the Ti_3SiC_2 grains exhibit obvious lamellar characteristic with the grain size in the range of several microns (Figure 3c), which is a typical microstructure feature of the reactively synthesized MAX phase [1–3].

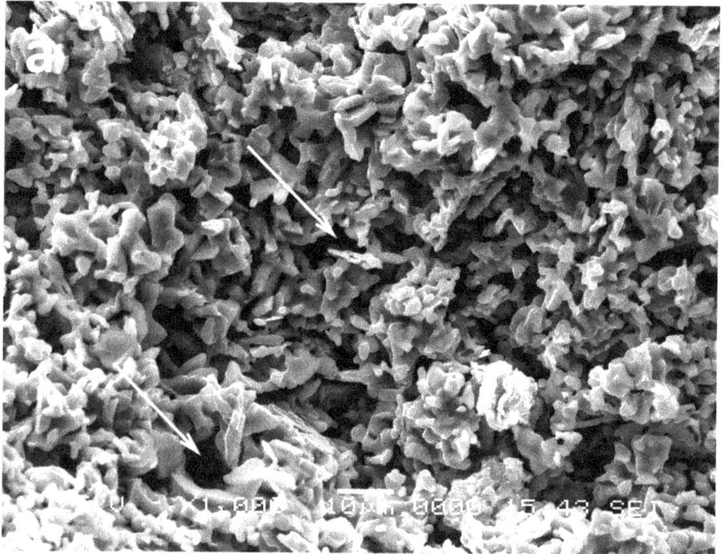

Figure 3. *Cont.*

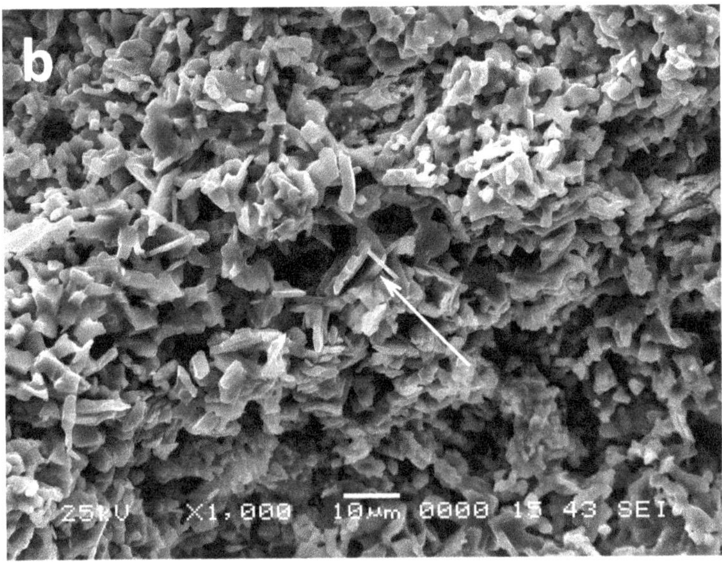

Figure 3. SEM images of the pore structure morphology of the sintered compacts with the average apertures of 6.2 μm (**a**) and 3.0 μm (**b**) and the lamellar microstructure of the synthesized porous Ti_3SiC_2 (**c**).

Figure 4 shows the relation between the flexural stress and strain for porous Ti_3SiC_2 with different average pore sizes. The variation rate of flexural stress with strain increases with decreasing the pore size. All the stress–strain curves show a typical elastic deformation stage directly followed by a fracture stage. However, the flexural stress increases nonlinearly with increasing strain, which is mainly attributed to the deformation of the MAX phase as a kinking nonlinear elastic solid through the formation of regular and incipient kink bands [12].

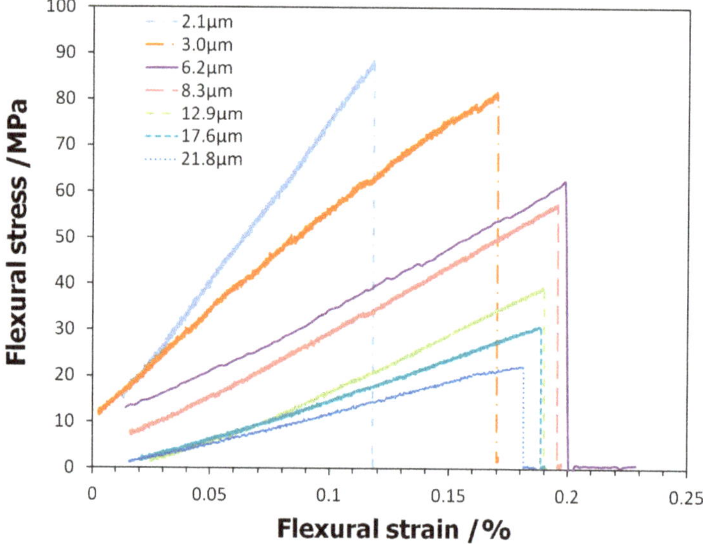

Figure 4. Flexural stress–strain curves of porous Ti_3SiC_2 with different average pore sizes.

Figure 5 shows the relation between the bending strength and the average pore size of porous Ti_3SiC_2. The bending strength increases in a nonlinear manner with decreasing the average pore size (Figure 5a). As the aperture decreases from 21.8 to 2.1 μm, the bending strength increases from 22 to 89 MPa. Through data fitting, the relationship between them is as follows:

$$\sigma = k \cdot d^{-\lambda} \tag{1}$$

where σ and d are the bending strength (MPa) and the average pore size (μm) of the porous material, respectively. k and λ are the constants with the values of 154.04 and 0.56, respectively. The determination coefficient R^2 of the fitting relation equation is 0.92. Given λ is close to 0.5, further relationship between σ and $d^{-0.5}$ is investigated, as shown in Figure 5b. It is observed that σ increases linearly with $d^{-0.5}$. Through data fitting, the relationship between them is as follows:

$$\sigma = \sigma_0 + k \cdot d^{-\frac{1}{2}} \tag{2}$$

where σ_0 and k are the constants with the values of 1.60 MPa and 135.48 MPa·$\mu m^{1/2}$, respectively.

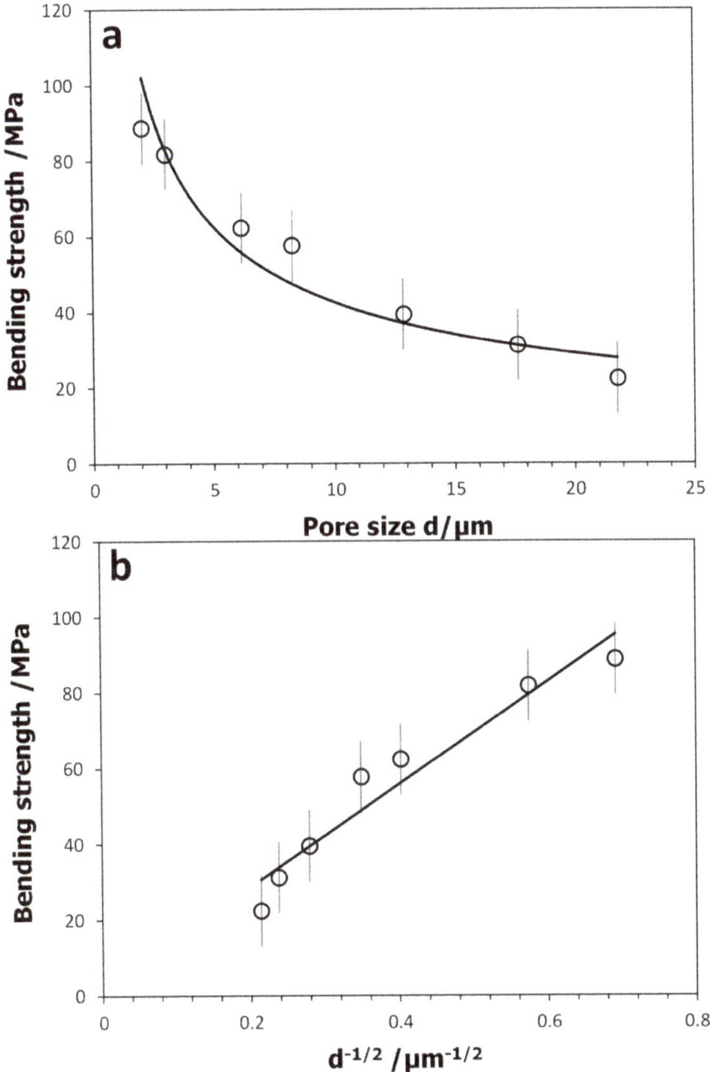

Figure 5. Relationship curves between the bending strength and the average pore size d (**a**) and $d^{-1/2}$ (**b**) for porous Ti_3SiC_2.

The determination coefficient R^2 of the fitting relation equation is 0.94, which shows a higher fitting degree than Equation (1). Equation (2) is the Hall–Petch equation reflecting the relationship between the microstructure parameter and the mechanical property [20,21]. The result indicates that the influence of the pore size on the strength of the reactively synthesized porous Ti_3SiC_2 compound can be described by the Hall–Petch equation under the condition of approximately the same porosities and phase purities. In fact, a similar linear relationship between strength and reciprocal square root of the pore size was also found in porous alumina ceramics [26]. As the main component of the porous material, the pore can be approximately regarded as a phase in composite material. The decrease of pore size weakens the size factor that causes stress concentration [27], which is conducive to the improvement of material strength. On the other hand, the content of MAX phase in the sintered compacts increases with decreasing the pore size (Figure 2). Compared with Ti_3SiC_2 phase, TiC impurity phase has higher hardness and brittleness. Under the action of external load, this kind of hard and brittle impurity phase can often cause the stress concentration, which results in the decrease of the strength of the material. Therefore, the synthesis of Ti_3SiC_2 phase with high purity has always been the focus of researchers [1,3,25].

In this work, the pore size of porous Ti_3SiC_2 compounds has been adjusted by changing the particle size of TiH_2 powder. In the reactive synthesis process, there is a strict linear relationship between the powder size of TiH_2 and the pore size [6]:

$$d_m = g \cdot d_p \tag{3}$$

where d_m and d_p are the maximum pore size (μm) of porous Ti_3SiC_2 and the powder particle size (μm) of TiH_2, respectively, and g is the proportionality coefficient. Under the same synthetic process conditions, the reduction of powder size is beneficial to obtain the resultant material with smaller pores [6].

Figure 6 shows the relationship between the flexural elastic modulus and the pore size. When the pore size decreases from 21.8 to 2.1 μm, the flexural modulus increases from 13 to 70 GPa, which increases rapidly with the further decrease of the pore size to several microns. For comparison, the Young's and the shear moduli of porous Ti_3SiC_2 with the porosity fraction of ca. 0.5 prepared by space holder were investigated to be about 50 GPa and 20 GPa, respectively [13]. Through data fitting, the relationship between them is as follows:

$$E = h \cdot d^j \tag{4}$$

where E and d are the flexural modulus (GPa) and the average pore size (μm) of porous Ti_3SiC_2, respectively. h and j are the constants with the values of 97.42 and −0.61, respectively.

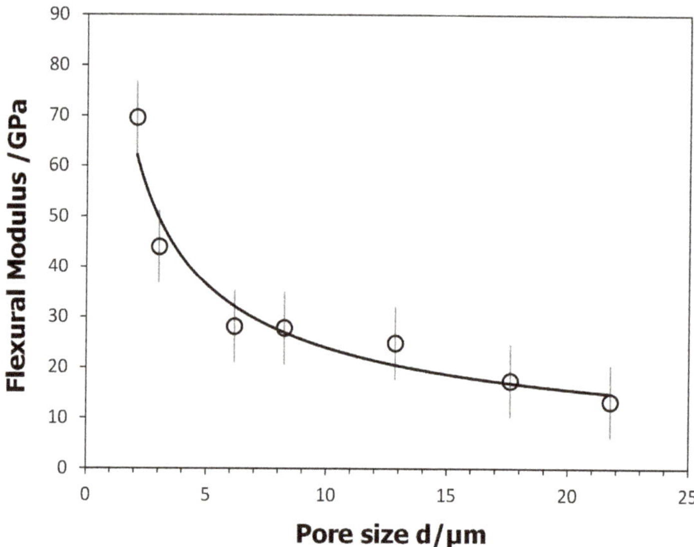

Figure 6. Relation curves between the flexural modulus and the average pore size for porous Ti_3SiC_2.

The determination coefficient R^2 of Equation (4) is 0.95, indicating that the flexural modulus of porous Ti_3SiC_2 has a relatively good power function relation with the pore size. In the investigations on the relation of pore size and elastic modulus for porous materials, there are two kinds of research results for different material types. On the one hand, it has been revealed that the elastic modulus increases slightly with the increase of the pore size for a porous alumina ceramic [26]. On the other hand, it has been found that the elastic modulus increases with the decrease of the pore size for a porous nickel metal [28]. In this study, the change behavior of Ti_3SiC_2 MAX phase is similar to that of the latter, which indicates that porous Ti_3SiC_2 exhibits similar characteristics to porous metals under the action of bending stress.

4. Conclusions

Reactively synthesized porous Ti_3SiC_2 compounds with different pore sizes have the characteristics of high purity of MAX phase and laminar microstructure. The flexural stress–strain curves of porous Ti_3SiC_2 compounds exhibit two stages of elastic deformation and fracture, and the elastic deformation behavior shows a characteristic of kinking nonlinear elastic solid. The bending strength increases with decreasing the pore size for the synthesized porous Ti_3SiC_2 with approximately the same porosities and phase purities, and the change law between them follows the Hall–Petch equation. The flexural modulus of porous Ti_3SiC_2 has a power function relationship with the pore size. The reduction of the pore size of porous Ti_3SiC_2 improves the phase purity, strength and stiffness while increasing its filtration accuracy.

Author Contributions: Conceptualization, Y.J. and Y.H.; methodology, Y.J.; validation, H.G.; formal analysis, X.L.; investigation, Y.J. and X.L.; resources, Y.J.; data curation, H.G.; writing—original draft preparation, Y.J.; writing—review and editing, Y.J. and Y.H.; supervision, Y.H.; project administration, Y.H.; funding acquisition, Y.J., X.L. and H.G. All authors have read and agreed to the published version of the manuscript.

Funding: This research was funded by the National Natural Science Foundation of China, grant number 51971251, 51774336, 51604305.

Acknowledgments: The work is financially supported by the National Natural Science Foundation of China (No. 51971251, 51774336, 51604305).

Conflicts of Interest: The authors declare no conflict of interest.

References

1. Liu, X.; Zhang, H.; Jiang, Y.; He, Y. Characterization and application of porous Ti_3SiC_2 ceramic prepared through reactive synthesis. *Mater. Des.* **2015**, *79*, 94–98. [CrossRef]
2. Wang, Z.; Jiang, Y.; He, Y. Oxidation behavior of reactively synthesized porous $Ti_3(Si,Al)C_2$ compound at 800 °C in ambient air. *Ceram. Int.* **2019**, *45*, 15482–15487. [CrossRef]
3. Liu, X.; Jiang, Y.; Zhang, H.; Yu, L.; Kang, J.; He, Y. Porous Ti_3SiC_2 fabricated by mixed elemental powders reactive synthesis. *J. Eur. Ceram. Soc.* **2015**, *35*, 1349–1353. [CrossRef]
4. Velasco, B.; Tsipas, S.A.; Ferrari, B.; Gordo, E. MAX phase foams produced via powder metallurgy process using water soluble space holder. *Powder Metall.* **2015**, *58*, 95–99. [CrossRef]
5. Velasco, B.; Gordo, E.; Tsipas, S.A. MAX phase Ti_2AlC foams using a leachable space-holder material. *J. Alloys Compd.* **2015**, *646*, 1036–1042. [CrossRef]
6. Liu, X.; Zhang, H.; Jiang, Y.; He, Y. Factors affecting the property of porous Ti_3SiC_2 metal ceramic fabricated through pressureless sintering. *J. Porous Mater.* **2015**, *22*, 1285–1290. [CrossRef]
7. Wang, Z.; Zhang, H.; Liu, X.; Jiang, Y.; Gao, H.; He, Y. Reactive synthesis of porous nanolaminate $Ti_3(Si,Al)C_2$ intermetallic compound. *Mater. Chem. Phys.* **2018**, *208*, 85–90. [CrossRef]
8. Sun, Z.M.; Murugaiah, A.; Zhen, T.; Zhou, A.; Barsoum, M.W. Microstructure and mechanical properties of porous Ti_3SiC_2. *Acta Mater.* **2005**, *53*, 4359–4366. [CrossRef]
9. Zhao, B.; Gain, A.K.; Ding, W.; Zhang, L.; Li, X.; Fu, Y. A review on metallic porous materials: Pore formation, mechanical properties, and their applications. *Int. J. Adv. Manuf. Technol.* **2018**, *95*, 2641–2659. [CrossRef]
10. Xiong, X.; Wang, Z.; Wang, X.; Liu, H.; Ma, Y. Enhancing the mechanical strength and air permeability of corundum porous materials using shape-modified coarse aggregates. *Ceram. Int.* **2019**, *45*, 11027–11031. [CrossRef]
11. Zhang, B.; Huang, H.; Lu, X. Fabrication and properties of C/SiC porous ceramics by grinding-mould pressing-sintering process. *J. Eur. Ceram. Soc.* **2019**, *39*, 1775–1780. [CrossRef]

12. Zhou, A.G.; Fraczkiewicz, M.; Barsoum, M.W. Mechanical damping in porous Ti_3SiC_2. *Acta Mater.* **2006**, *54*, 5261–5270.
13. Velasco, B.; Gordo, E.; Hu, L.; Radovic, M.; Tsipas, S.A. Influence of porosity on elastic properties of Ti_2AlC and Ti_3SiC_2 MAX phase foams. *J. Alloys Compd.* **2018**, *764*, 24–35. [CrossRef]
14. Finkel, P.; Zhou, A.G.; Basu, S.; Yeheskel, O.; Barsoum, M.W. Direct observation of nonlinear acoustoelastic hysteresis in kinking nonlinear elastic solids. *Appl. Phys. Lett.* **2009**, *94*, 241904. [CrossRef]
15. Zhou, A.G.; Basu, S.; Friedman, G.; Finkel, P.; Yeheskel, O.; Barsoum, M.W. Hysteresis in kinking nonlinear elastic solids and the Preisach-Mayergoyz model. *Phys. Rev. B* **2010**, *82*, 094105. [CrossRef]
16. Liu, X.; Zhang, Q.; Zhang, H.; Jiang, Y.; He, Y. Development and characterization of microporous Ti_3SiC_2 ceramic membranes for filtration of microorganisms. *J. Mater. Sci.* **2016**, *51*, 2594–2597. [CrossRef]
17. Gao, H.Y.; He, Y.H.; Zou, J.; Shen, P.Z.; Jiang, Y.; Liu, C.T. Mechanical properties of porous Fe-Al intermetallics. *Powder Metall.* **2015**, *58*, 197–201. [CrossRef]
18. Liu, G.; Lu, D.H.; Liu, X.W.; Liu, F.C.; Yang, Q.; Du, H.; Hu, Q.; Fan, Z.T. Solute segregation effect on grain boundary migration and Hall–Petch relationship in CrMnFeCoNi high-entropy alloy. *Mater. Sci. Technol.* **2019**, *35*, 500–508. [CrossRef]
19. Ryou, H.; Drazin, J.W.; Wahl, K.J.; Qadri, S.B.; Gorzkowski, E.P.; Feigelson, B.N.; Wollmershauser, J.A. Below the hall–petch limit in nanocrystalline ceramics. *ACS Nano* **2018**, *12*, 3083–3094. [CrossRef]
20. Wollmershauser, J.A.; Feigelson, B.N.; Gorzkowski, E.P.; Ellis, C.T.; Goswami, R.; Qadri, S.B.; Tischler, J.G.; Kub, F.J.; Everett, R.K. An extended hardness limit in bulk nanoceramics. *Acta Mater.* **2014**, *69*, 9–16. [CrossRef]
21. Wang, C.; Jiang, C.; Cai, F.; Zhao, Y.; Zhu, K.; Chai, Z. Effect of shot peening on the residual stresses and microstructure of tungsten cemented carbide. *Mater. Des.* **2016**, *95*, 159–164. [CrossRef]
22. Roa, J.J.; Jiménez-Piqué, E.; Tarragó, J.M.; Sandoval, D.A.; Mateo, A.; Fair, J.; Llanes, L. Hall–Petch strengthening of the constrained metallic binder in WC–Co cemented carbides: Experimental assessment by means of massive nanoindentation and statistical analysis. *Mater. Sci. Eng. A* **2016**, *676*, 487–491. [CrossRef]
23. Li, C.L.; Mei, Q.S.; Li, J.Y.; Chen, F.; Ma, Y.; Mei, X.M. Hall–Petch relations and strengthening of Al-ZnO composites in view of grain size relative to interparticle spacing. *Scr. Mater.* **2018**, *153*, 27–30. [CrossRef]
24. He, W.; Weng, Q.G.; He, Y.H.; Jiang, Y. Preparation of ultrafine Ti powder by inhibitor coated/HDH combined method. *Powder Metall.* **2013**, *56*, 239–244. [CrossRef]
25. Zhang, Z.F.; Sun, Z.M.; Hashimoto, H. Rapid synthesis of ternary carbide Ti_3SiC_2 through pulse-discharge sintering technique from Ti/Si/TiC powders. *Metall. Mater. Trans. A* **2002**, *33*, 3321–3328. [CrossRef]
26. Ashizuka, M.; Ishida, E.; Matsushita, T.; Hisanaga, M. Elastic modulus, strength and fracture toughness of alumina ceramics containing pores. *J. Ceram. Soc. Jpn.* **2002**, *110*, 554–559. [CrossRef]
27. Xu, Z.; Wen, W.; Zhai, T. Effects of pore position in depth on stress/strain concentration and fatigue crack initiation. *Metall. Mater. Trans. A* **2012**, *43*, 2763–2770. [CrossRef]
28. Li, Z.; Yang, L.; Li, Y.; Yang, Y.; Zhou, C.; Ding, Y.; Zhao, J.; Li, Y. Effects of pore size on the mechanical properties of three-dimensionally ordered macroporous nickel. *Mater. Des.* **2013**, *45*, 52–55. [CrossRef]

Article

Experimental Study on Zn-Doped Al-Rich Alloys for Fast on-Board Hydrogen Production

Dan Liu [1], Qian Gao [1,*], Qi An [1], Hongchao Wang [1], Jilun Wei [2] and Cundi Wei [1,*]

1 Key Laboratory of Automobile Materials, Ministry of Education, College of Materials Science and Engineering, Jilin University, Changchun 130025, China; liud14@mails.jlu.edu.cn (D.L.); ann_940115@163.com (Q.A.); whc19@mails.jlu.edu.cn (H.W.)

2 The College of Liberal Arts and Science, Arizona State University, AZ 85287, USA; w790896570@outlook.com

* Correspondence: gaoqian@jlu.edu.cn (Q.G.); weicd@jlu.edu.cn (C.W.)

Received: 31 January 2020; Accepted: 28 February 2020; Published: 3 March 2020

Abstract: For the purpose of investigating the effect of Zn replacement of In_3Sn on the hydrogen production performance of Al-rich alloy ingots, Al-Ga-In_3Sn alloys with various Zn dosages (0–5 wt.%) were prepared by a traditional melting and casting technique. The phase compositions and microstructures were characterized using X-ray diffractometer (XRD) and scanning electron microscope (SEM) with an Energy Dispersed X-ray system (EDS). The SEM results indicate that, with a small amount of Zn instead of In_3Sn, the number and total area of grain boundary (GB) phases will decrease gradually, and the average single GB area will eventually stabilize. The distribution of Zn in the alloy is similar to that of Ga, and an area with high Zn content appeared in the high-Zn-doped sample. The melting behaviors of Al with other metals were measured by DSC. The reaction of these alloys and water were investigated at different temperatures. Compared with Al-Ga-In_3Sn alloy, low addition of Zn changed the composition of GB phase and increased the maximum hydrogen production rate. The reason for the changes in the hydrolysis reaction of Al with the addition of Zn was discussed.

Keywords: Al-rich alloy; intermetallic compound; hydrogen generation; Zn addition

1. Introduction

Due to the limited world proven reserves and environmental degradation caused by fossil fuel consumption, it is clear that social development based on traditional energy sources is not sustainable. Consequently, it has been an irresistible trend to find alternate green energy which appeals for a clean, co-development and efficient energy future [1–3]. Hydrogen as an energy carrier has a high energy density about three times higher than gasoline. Furthermore, the water produced after hydrogen energy combustion is absolutely clean and will not cause environmental pollution. In the past few decades, the main approaches to produce hydrogen could be grouped into the following categories [4–7]: water electrolysis, fossil fuel gasification, chemical processing techniques and biological methods. So far, a wide range of applications of hydrogen has not been realized through these methods. Many problems need to be solved, such as expensive cost, poor conversion efficiency, insecurity in hydrogen storage and transportation. In this case, it is extremely essential to produce hydrogen in situ, particularly in the aspects of emergency power provided in disaster-stricken areas, portable electronic equipment and on board vehicles. Hence, researchers are constantly looking for new ways of producing hydrogen gas.

In recent years, an emerging in situ method through the hydrolysis of Al alloys has been increasingly attracting the interest of researchers [8–10]. The hydrogen generated by such a method has many advantages. Firstly, the hydrogen can be produced almost anywhere, which eliminates the strict requirements for storage and transport. Secondly, the generated hydrogen has high purity and hence, can be directly supplied to internal combustion engines or fuel cells. Most of all, aluminum as a metal is rich in the earth's crust; it also has a high hydrogen generation capacity. Through theoretical calculations, the amount of hydrogen produced per 1 g of Al can reach 1.244 L. Nevertheless, a fine and close oxide film easily forms on the pure Al surface which impedes the subsequent reaction of Al hydrolysis. Several treatment measures have been taken to activate Al as follows [11–14]: alkali aqueous activation, nano/micro-crystallization of Al, and mixing particular metallic oxide or mineral salts. Nevertheless, these processing ways have certain disadvantages, e.g., acid or alkaline solution is extremely corrosive and the pre-preparation work of milling is time consuming. Adding some low melting alloys to Al and alloying was a fascinating method of activation [15–18]. This approach can produce Al alloys in large quantities in a short time when required, and Al can be hydrolyzed at a mild temperature to generate hydrogen.

Previous studies have shown that the size and composition of GB phase play the dominant role in the process of hydrolysis reaction between Al and water [19,20]. The GB phase containing low melting point metals can prevent the oxidation of Al. Meanwhile, these activated areas also provide a transmission channel to make it easier for Al to enter the reaction site. Up to now, the Al-Ga-In-Sn alloy has been investigated in a systematic study [21,22]. The melting point of liquid Ga-In-Sn phase is 10.4 °C, which has a great correlation with the reaction temperature. In addition, In and Sn mainly exist in two kinds of interstitial phases: In_3Sn (β) and $InSn_4$ (γ). When the content of In_3Sn is high, the alloy exhibits high energy transduction efficiency and fast hydrogen generation rate. Yet, compared with Al-Ga-$InSn_4$, Al-Ga-In_3Sn alloy also suffers from some disadvantages. Indium is expensive and hard to recycle. Furthermore, the surface of this alloy easily reacts with the moisture in the air, which requires higher storage conditions. For this reason, in consideration of cost, a fifth metal needs to be introduced to further improve the comprehensive performance of the Al-Ga-In_3Sn alloy.

Combined with results of previous studies, various dopants have different effects on the Al-Ga-In-Sn alloy. When Bi [23] or Mg [24] are added to these Al-based alloys, new GB phases (InBi or Mg_2Sn) are formed, and the alloy exhibits different energy conversion efficiency. With the addition of the grain refiners (Ti, Al_2O_3, $AlTi_5B$ etc.) [25–28], the grain size and morphology of the alloys have changed significantly, which also affects hydrogen production performance. In a word, the effect of doped metals on the alloy is usually to change the GB composition or grain size.

It is acknowledged that Zn is an important metal used in industry. Studying the doping of Zn in Al-Ga-based hydrogen-producing alloys will help a wider choice of raw materials and the use of waste Al in the future. In order to achieve practical application, the process of hydrogen production needs to be controlled in an on-board hydrogen supply unit. Thus, it is worth undertaking a comprehensive investigation into the specific influence of Zn in water splitting reactions. By adjusting the mass ratio of In_3Sn and Zn, alloys with different hydrogen production properties can be obtained. Differentiated reaction temperature and start-up time help to realize the control of an on-board hydrogen supply.

For the sake of high economic benefits and excellent hydrogen production performance, Zn was introduced into Al-Ga-In_3Sn alloys to reduce the contents of In and Sn. The micrographs and crystalline structure of alloys were examined by XRD, SEM with EDS. Then, the alloy melting behavior was measured by DSC and corresponding hydrogen production performance was discussed. In addition, we used an isothermal kinetic model to calculate the activation energy of alloy hydrolysis and discussed the relevant reaction mechanisms.

2. Materials and Methods

In order to explore the relationship between physicochemical properties and material properties of Al-Ga-In_3Sn-Zn, a series of alloys with different Zn contents were prepared and some experimental tests were performed. Zn was hypothesized to change the composition of the GB phase and thus affect the hydrogen production performance. In previous studies, the eutectic reaction in the GB phase was identified to be a key factor in the hydrolysis of Al. The continuous replacement of In and Sn, the main component of the GB phase, by Al can help verify this theory and determine other possible reaction mechanisms.

2.1. Preparation of Materials

Six kinds of Alloy ingots (Table 1) were prepared using a simple melting and casting technique. The weights of the cast ingots were kept at 20 g approximately. In these samples, the contents of Al and Ga remain fixed, the mole ratio of In and Sn remained a constant of 3:1, while Zn content changed from 0 wt.% to 5 wt.%.

Table 1. Compositions of prepared alloys (wt.%).

No.	Al	Ga	In	Sn	Zn
1	90	2.5	5.58	1.92	0
2	90	2.5	4.83	1.67	1
3	90	2.5	4.09	1.41	2
4	90	2.5	3.35	1.15	3
5	90	2.5	2.60	0.90	4
6	90	2.5	1.86	0.64	5

The starting mixture containing Al (purity, 99%) and dopants (Ga, In, Sn, and Zn of at least reagent grade) was heated at a rate of 10 °C/min up to 800 °C and held at this temperature for 1 h. The smelting was protected by nitrogen in the furnace. After that, the molten metals were stirred at a constant speed for 10 min by a stirring paddle and then cast into the mold with a cylindrical recess and cooled in air. Finally, the cooled ingot was wrapped with a sealing film and stored in a vacuum sealed bag for the follow-up experiments.

2.2. Characterization

For the purpose of determining the phase compositions, the sections of casted ingots were polished and then studied by an X-ray diffractometer (XRD) employing a DX-2700 X-ray diffractometer (Dandong Fangyuan, China) with monochromatic $Cu_{K\alpha}$ radiation ($\lambda_{K\alpha1}$ = 0.154056 nm). The detection angle (2θ) were collected from 30° to 80° at a scanning speed of 0.03°/step. The morphology and element analysis were characterized by a cold field scanning electron microscope (SEM, HITACHI SU8020) equipped with an Energy Dispersed X-ray (EDS, Bruker QUANTAX 200) system. In order to minimize the oxidation of fracture surface, the specimens were put into the chamber at once after they were made. The melting behaviors of samples were detected by differential scanning calorimeter (DSC, PERKIN-ELMER DSC 7). Indium was used as a reference substance to calibrate the apparatus by testing its melting point. Each sample of 40 mg or so was put into Al pan under a flowing nitrogen atmosphere. In order to observe low-temperature phase transitions in the sample, the related data were collected from −20 to 225 °C at an invariable rate of 10 °C/min.

2.3. Measurement of Hydrogen Production

The apparatus was used for hydrogen yield measurement has been illustrated in detail in our previous research [29]. Each sample was cut into a cylindrical sheet with a diameter of about 15 mm and a weight of 0.5 g. A 250 mL glass container was placed in a thermostatic water bath. When 100 mL deionized water reached the predetermined temperature, the plug was closed immediately as soon as a pre-prepared sample was placed into water. Considering that water vapor may cause errors in experimental results, as an improvement, a condensation tube was installed on the outlet side of the Pyrex glass reactor, and the time taken by the reaction for every 10 mL of hydrogen produced was recorded. In the meantime, the height of the calibration bottle was continuously adjusted to ensure that its water level was consistent with the glass burette. The data of hydrogen release rate and reaction time were recorded automatically by the mass flow meter (Alicat Scientific) and stored in a computer. According to the instrument operating procedures, a dryer was directly connected between the mass flow meter and the reactor. In the experiments, theoretical hydrogen production volume was calculated according to an ideal gas formula (PV = nRT). Under standard conditions (273 K, 10^5 Pa), 1 g Al can generate 1.244 L of hydrogen.

3. Results

3.1. Hydrogen Generation Performance

The alloys' hydrogen production yields at 40 °C are shown in Figure 1. The volumes of hydrogen generated from the first four samples (0–3 wt.% Zn) approach the theoretical value, implying that the alloys could hydrolyze thoroughly. As the sample was put into water, they split into small pieces at once and hydrogen bubbles started to pop up. However, the time used for completing the hydrolysis reaction is quite different. The samples containing 0 and 3 wt.% Zn have the shortest reaction time, while the sample of 2 wt.% Zn lasted tens of minutes.

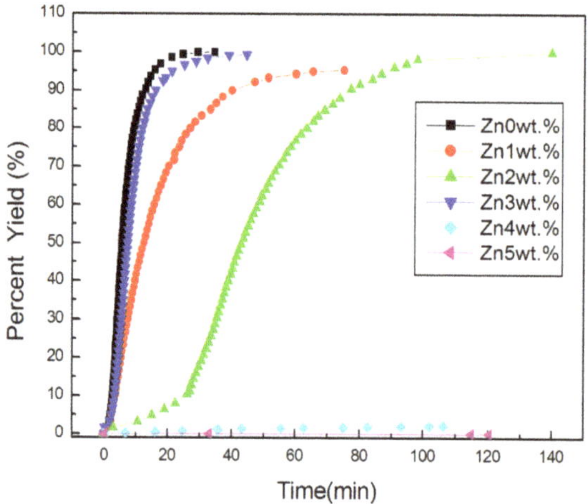

Figure 1. Hydrogen yields of the Al-Ga-In_3Sn-Zn alloys at temperatures of 40 °C.

Figure 2 shows the hydrogen production rate curves of the Al-Ga-In_3Sn-Zn alloys at different water temperature. It can be clearly seen that the hydrogen production performance is closely related to the reaction temperature. As the reaction temperature rises, the reaction time is shortened and the reaction rate increased markedly.

3.2. Characterization of Al Alloys

3.2.1. XRD Analysis

Figure 3 shows the XRD patterns of the synthesized alloys containing various Zn content. It can be seen that all samples mainly consist of two phases: crystalline Al phase (JCPDS file #04-0787) and In_3Sn intermetallic compound (JCPDS file #07-0345). Besides, Ga-related diffraction peaks have not be observed. Similar phenomenon is common in previous studies of Al-Ga-based alloys, and it has been confirmed that Ga exists as a form of Al (Ga) solid solution [30–32]. Yet, due to a low doping amount of Ga in the raw materials and the close atomic radius of the two atoms, the positions of Al diffraction peaks have not changed significantly. Additionally, no significant characteristic peaks of crystalline Zn phase were found, which maybe owing to the low doping amount or similar properties of Zn and Ga in Al (i.e., high solid solubility, similar atomic radius). Thus, the existence of Zn needs to be further confirmed by EDX and other analysis.

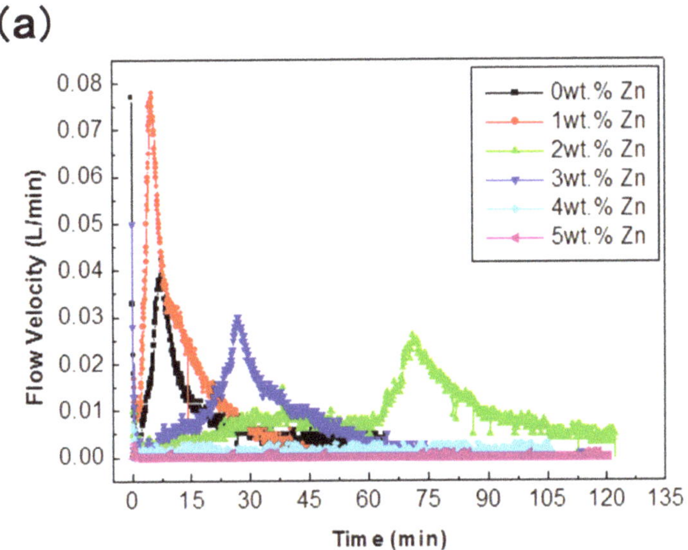

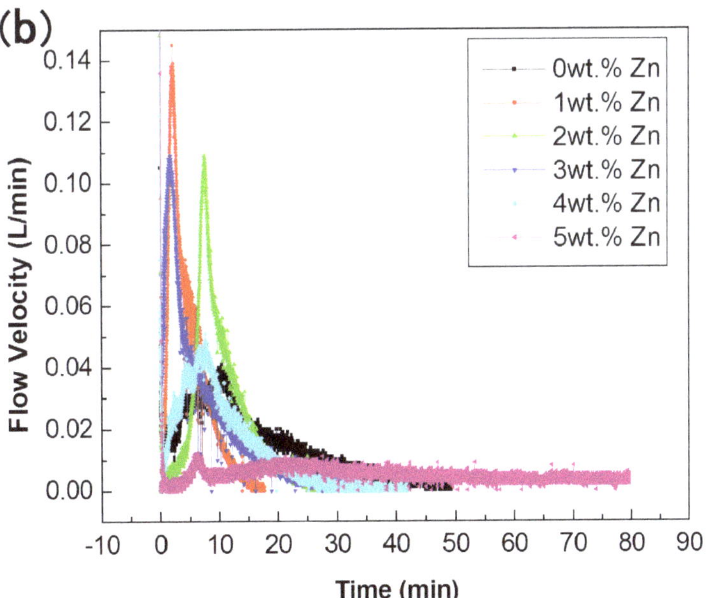

Figure 2. *Cont.*

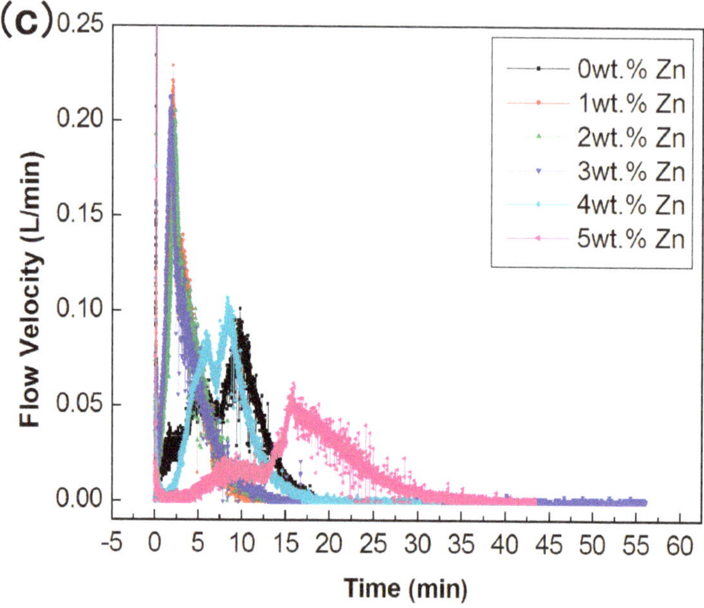

Figure 2. Hydrogen generation rates of the Al-Ga-In_3Sn-Zn alloys at temperatures of (**a**) 40 °C, (**b**) 50 °C, (**c**) 60 °C.

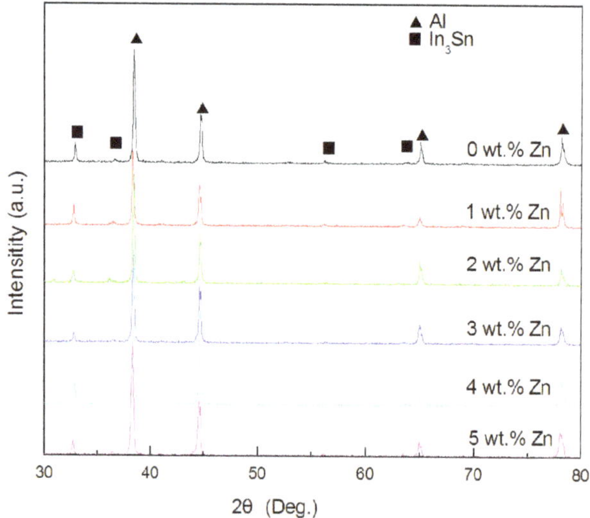

Figure 3. XRD patterns of the remnant alloys.

3.2.2. SEM Observation

The SEM images showed in Figure 4 are the fracture surfaces of Al-Ga-In_3Sn-Zn alloys with various content of Zn. All the Al grains in the prepared ingots present columnar shapes and directional growth during solidification. The Al grain sizes (column widths) range from 20 to 50 μm. According to statistics, the average size is approximately 35 μm, and the standard deviation is 6μm, which is similar to the Zn-free alloy (33 ± 5μm). Zn as an additive, which is similar to Bi [23,29], affects the grain size slightly, but it is completely different from Ti [4], and the segregation has a greater effect on grain size [33–35]. Furthermore, the columnar surfaces are covered by a lot of small granular and irregular particles. Considering the solidification process, the GB phase begins to nucleate and grow at the grain boundaries, which is different from the nucleation time of Al grains. From the SEM images (Figure 4a–c), several cracks exit on the grain surfaces; these alloys easily oxidize in the air and lose their metallic luster. On the other hand, the grains with high Zn content (Figure 4d–f) are much more complete. This will delay the arrival of the fast hydrogen generation stage, because the alloy needs to split into small pieces to enable rapid hydrolysis.

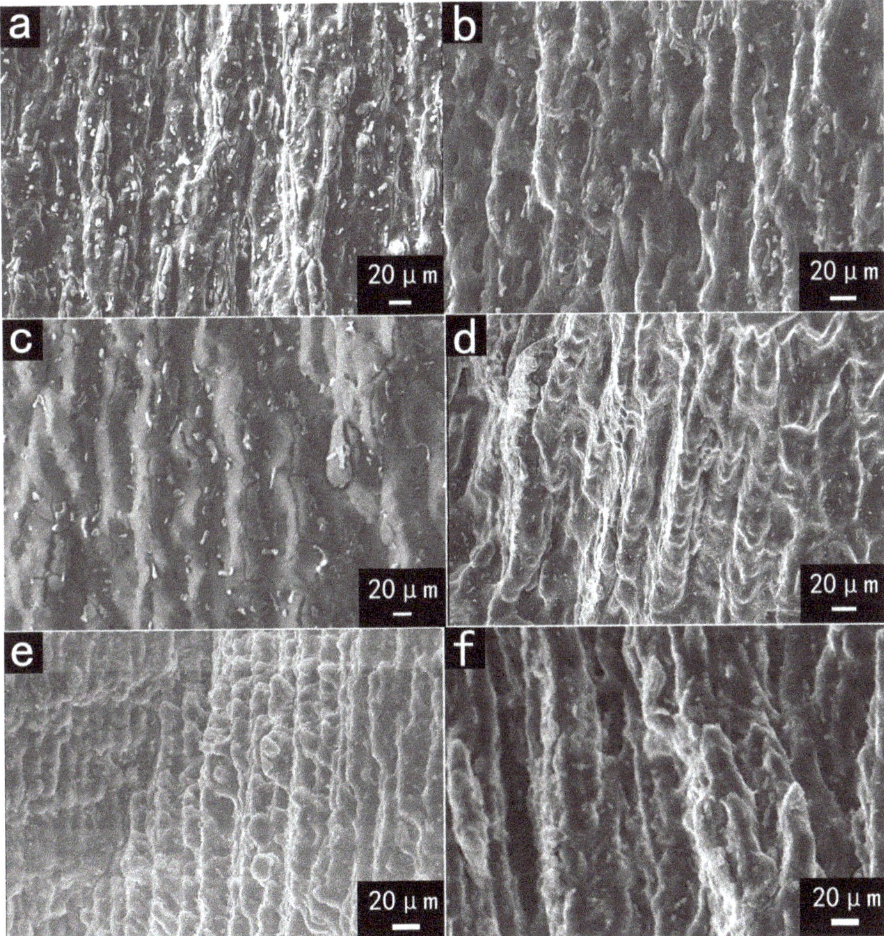

Figure 4. SEM images of Al-rich alloys containing (**a**) 0 wt.% Zn, (**b**) 1 wt.% Zn, (**c**) 2 wt.% Zn, (**d**) 3 wt.% Zn, (**e**) 4 wt.% Zn, and (**f**) 5 wt.% Zn.

W. Wang et al. [19] had indicated that the changes in size and number of GB particles will lead to different hydrogen production performance of alloys. In their study, the larger number and size of particles contributed to increasing the rate of hydrogen production. A statistical analysis of grain-boundary phase particle areas, obtained from 40 images, is shown in Figure 5. As the doping amount of Zn is within the range of 0–5 wt.%, total area of GB particles in the observed area decreases rapidly (from 6.5% to 2.5%). The average area of individual GB particles has also changed and was reduced from 15 to 12 μm. When the Zn content exceeded 2 wt.%, the particle size was stable (between 11 and 12 μm). The results indicate that the relative content of Zn and In_3Sn has a certain effect on the formation of GB phase, and these phenomena need to be further explained in conjunction with EDX data.

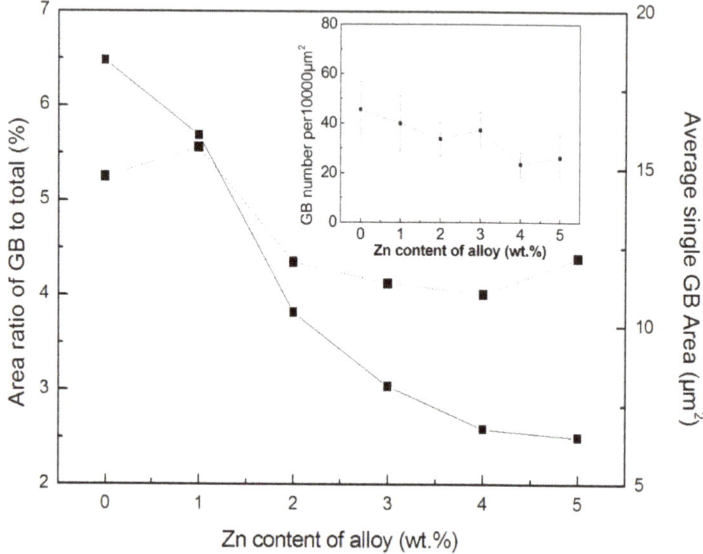

Figure 5. Area ratio of GB particles to the total area and GB area versus Zn contents. Inset shows the GB particle numbers versus Zn content.

3.2.3. EDX Analysis

Since the EDX data are collected from the fracture surface of the sample, the relevant data are only used as references for the distribution law of each element. EDX analysis from Al-Ga-In_3Sn-Zn ingots was performed to identify the element distribution in Al grains (G) and GB phases. SEM images of GB phase particles are illustrated in Figure 6, and the EDX data are listed in Table 2. It can be seen that the samples include a lot of O, this is owing to the rapid fracture surface oxidation when the sample is exposed to air during preparation. The elements of Al grains mainly include Al, Ga and Zn (if doped) besides O. Since Ga and Zn have high solid solubility in Al and an approximate atomic radius, it is easy for them to get into the Al crystal lattice. A small amount of In and Sn, found in the matrix, were quenched into Al grain due to a quick cooling rate during solidification. In Zones 2, 4, 6 and 8, the atomic ratio of In:Sn in these GB particles approaches 3:1, which is very close to the mole ratio of the two in the raw material during the preparation process. Meanwhile, the solid solubility of In and Sn in Al grains is particularly low. When the content of Zn increases, as shown in Figure 6e, a small amount of Zn is found almost in its particulate form (91.7 at.% Zn), which is dispersed in the alloy ingots.

Zn exists in both the G and GB phases. The typical characteristic of the GB phase is the high content of In and Sn elements. This explains why the reduction of In_3Sn content in Figure 5 led to a relative decrease in the total area and number of GB phases. However, with an equivalent reduction in the amount of In_3Sn, the variation law of the total area and single area of the GB phase (the slope of the solid line and the dotted line in Figure 5) are different. This shows that when a small amount of Zn is doped, Zn mainly forms a solid solution with Al, causing the GB phase area to decrease rapidly. When more Zn is added, Zn starts to enter the GB phase, so the decline rate of the solid line which shows the total area of the GB phase slows down, and the area of a single GB phase is basically unchanged, staying around 11–12 μm.

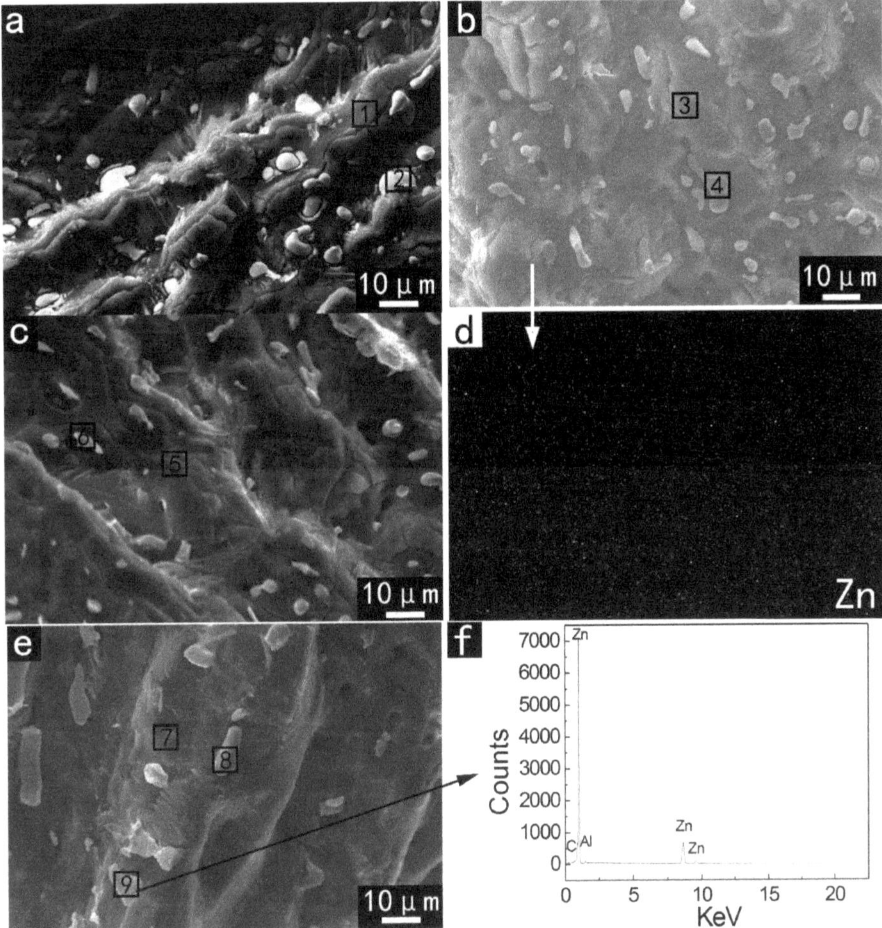

Figure 6. SEM images from fracture surface of Al-rich alloys, (**a**) 0 wt.% Zn, (**b**) 1 wt.% Zn, (**c**) 3 wt.% Zn and (**e**) 5 wt.% Zn; (**d**) EDX mapping scan spectrum of (**b**); (**f**) typical EDX mapping of Zone 9.

Table 2. Composition of the Al-Ga-In_3Sn-Zn alloys (In brackets are the data after oxygen removal).

Zone	Phase	Element (at.%)					
		Al	Ga	In	Sn	Zn	O
1	G	47.44(98.18)	0.72(1.49)	0.08(0.17)	0.08(0.17)	-	51.68
2	GB	12.81(32.65)	0.82(2.09)	19.22(48.98)	6.39(16.28)	-	60.76
3	G	38.90(97.71)	0.55(1.38)	0.08(0.20)	0.11(0.28)	0.16(0.40)	60.19
4	GB	11.82(22.38)	1.27(2.40)	30.02(56.83)	9.50(17.99)	0.22(0.42)	47.18
5	G	47.40(97.15)	0.57(1.17)	0.28(0.57)	0.06(0.12)	0.47(0.96)	51.21
6	GB	13.96(20.03)	1.67(2.40)	43.72(62.74)	9.42(13.52)	0.91(1.31)	30.32
7	G	68.46(95.99)	1.13(1.58)	0.17(0.24)	0.02(0.03)	1.54(2.16)	28.68
8	GB	3.09(3.58)	3.14(3.64)	61.03(70.78)	18.24(21.15)	0.73(0.85)	13.77
9	High-Zn region	4.71(4.85)	0.58(0.60)	-	0.20(0.21)	91.70(94.37)	2.83

3.2.4. DSC Curves

Figures 7 and 8 show the typical DSC curves of Al-rich alloys. One wide endothermal peak or exdothermal peak emerges in the DSC trace. The peak temperature is about 120 °C, which is supposed to be due to the eutectic reaction of Al and In_3Sn according to the In-Sn phase diagram (75 at.% In-25 at.% Sn). Compared with Al-Ga-In_3Sn alloy, the diffusive peak moves to the cold side when a small amount of Zn is doped, and then moves to the opposite side slightly accompanied by the increase of Zn doping.

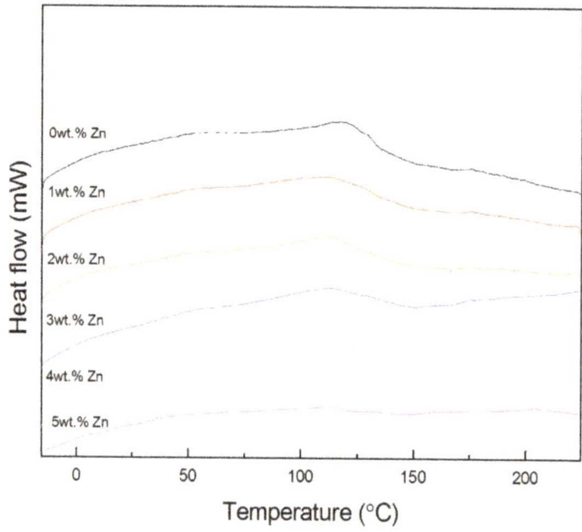

Figure 7. DSC curves of the Al-Ga-In_3Sn-Zn alloys tested by heating cycles.

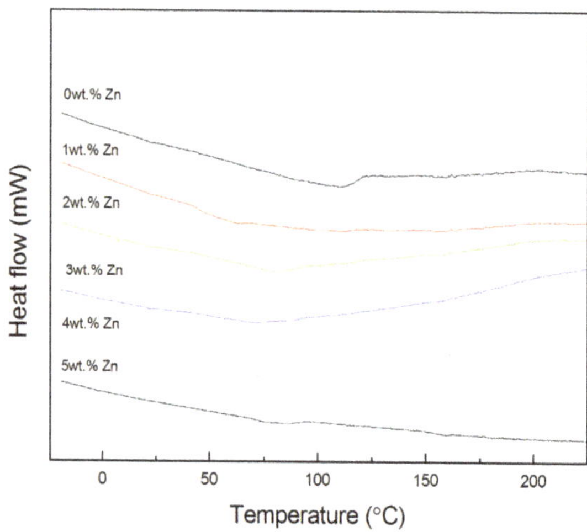

Figure 8. DSC curves of the Al-Ga-In_3Sn-Zn alloys tested by cooling cycles.

3.3. *Kinetic Parameters*

The formulas used to calculate the kinetic parameters of the Al hydrolysis reaction have been introduced in previous studies [36,37]. The activation energy (Ea) has been calculated (Figure 9) by the Arrhenius diagram of the hydrogen generation yields derived from various temperatures (Figure 10).

When the Zn doping amount is increased from 1 to 5 wt.%, the activation energy of these alloys increases accordingly from 59 to 139 kJ/mol.

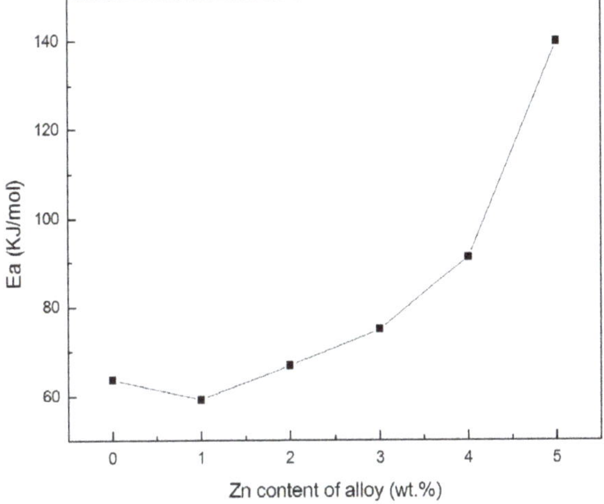

Figure 9. The activation energy versus Zn content.

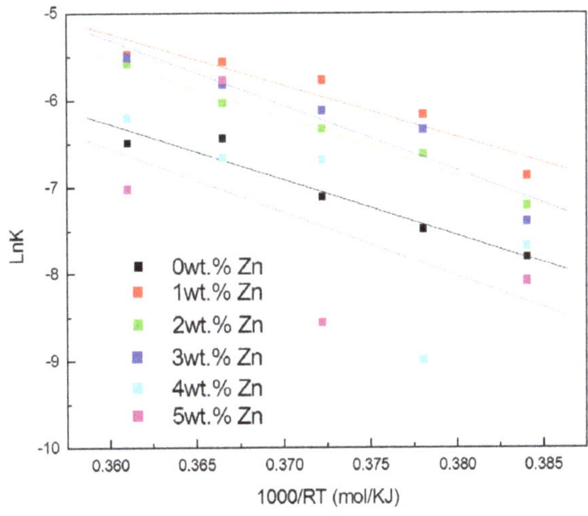

Figure 10. Arrhenius plots for the isothermal reaction of alloy hydrolysis.

From previous studies, the factors affecting the activation energy are as follows:

(1) W. Wang and coworkers [30] found that the Ea of Al-Ga-In-Sn alloys ranged from 53 to 77 kJ/mol. The researchers believe that this change is the result of different grain size (increases from 23 to 258 μm). Our results of the Zn-free alloy (Al grain size: 33 ± 5 μm; Ea: 63 KJ/mol) are very close to their values (37 ± 16 μm, 56 ± 5 KJ/mol).

(2) Jeffrey T. Ziebarth and coworkers [36] speculated that the compositions might affect the activation energy. The eutectic reaction between GB phase and Al is identified to be a key factor in alloy hydrolysis. The addition of Zn changes the GB phase from a ternary alloy (Ga-In-Sn) to a quaternary alloy (Ga-In-Sn-Zn). Adding a small amount of Zn means that the relative contents of In and Sn are reduced, and the Ga-In-Sn mass ratio is closer to the eutectic point (Ga-In-Sn Ternary phase diagram),

but as more low-melting metals (In_3Sn) are replaced by Zn (more than 1 wt.%), the energy required for the reaction increases dramatically. Hence, it makes sense that the activation energy varies with chemical composition.

(3) The segregation in alloys may also cause an increase in the activation energy [35]. A Zn-rich particle was found in the high Zn alloy, which is most likely due to segregation. Although the number of these particles is very small, the degree of influence of these segregations on the hydrolytic activation energy needs further study.

4. Discussion

According to the previous experimental results, it is found that the hydrogen production performance of alloys with different Zn doping amounts is significantly different. The composition of the material directly affects its physical and chemical characteristics. Therefore, the determination of the main phases of the alloy and the qualitative analysis of the distribution characteristics of Zn should be discussed first.

In addition to the Al and In_3Sn phases determined by XRD testing, the existence of Ga as a solid solution of Al (Ga) has also been verified in many studies [30–32]. The EDX analysis shows that Zn exists in both the G and GB phases. As shown in Figure 6d and EDX mapping of the whole image in Figure 6b, Zn atoms are randomly dispersed throughout the entire region, which is similar to the distribution of Ga in Al alloys [12]. It is further proven that Zn is more likely to exist as a solid solution combined with the XRD analysis and Al-Zn phase diagram. In addition, a small part of Zn will appear on the grain surface in the form of Zn-rich particles.

The entry of Zn into the GB phase means that the eutectic reaction of Al and GB phases will change accordingly, which is identified to be a key factor affecting the hydrolysis reaction of Al. The analysis results of DSC and activation energy confirmed this point, and their change rules are highly consistent, which can be explained by the same reason: As Zn begins to entered into the GB phase, the main composition of GB phase was changed from Ga-In-Sn to Ga-In-Sn-Zn, and the eutectic temperature and Ea of the grain boundary phase decreased. On the other hand, in the process of Zn taking the place of In_3Sn, the melting point of Zn is significantly above other GB elements, leading to the eutectic temperature and Ea increases gradually. In addition, high activation energy will effectively increase the difficulty of the alloy to react with moisture in the air, which is conducive to long-term storage and long-distance transportation. Therefore, this can also be considered as a criterion for judging the practical value of alloys.

However, this mechanism does not perfectly explain that the sample (3 wt.% Zn) still has excellent hydrogen production performance. According to the conclusion of the previous study [37,38], two main mechanisms can be used to discuss the occurrence of these phenomena. On the one hand, the eutectic reaction of Al and GB phase is the origin of the Al hydrolysis reaction. When Zn was added, the sample became less susceptible to split when exposed to water. At the same time, the area of a single GB phase was reduced to a minimum value at 2 wt.% Zn (Figure 5), which delayed the startup time of the reaction, and further delayed the arrival of the maximum hydrogen production rate (Figure 11). On the other hand, the Zn will form different micro batteries with other metals and Al corrodes rapidly as the electrode potential of Al–Zn is formed. It is clear from Figures 2a,b and 11 that the time consumed by the hydrolysis reaction of 2 wt.% Zn alloy is greater than the 3 wt.% one. Although the single GB area of the two is similar (Figure 5), the high Zn content means that more Al-Zn micro-galvanic cells will promote the reaction. Nevertheless, the content of Ga, In and Sn in the GB phase directly determines the starting temperature of the hydrolysis reaction. That is why the high-Zn content alloys (4 wt.%, 5 wt.% Zn) cannot hydrolyzed at a lower reaction temperature of 40 °C and low content of In_3Sn makes the alloy stay intact in water.

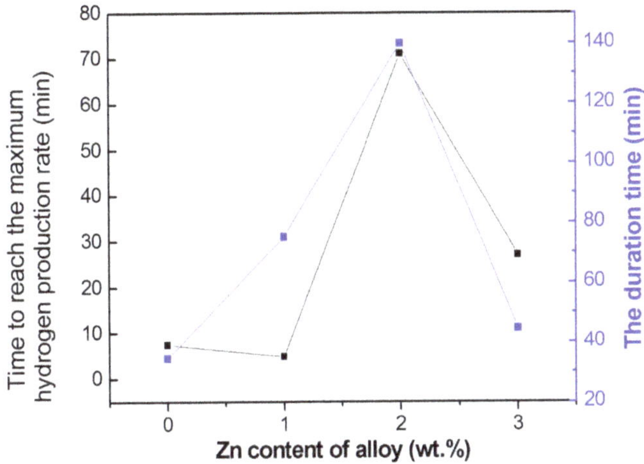

Figure 11. Statistics on hydrogen production time of Al alloys at 40 ° C.

As shown in Figure 12, when the temperature increased from 40 to 60 °C, the max hydrogen production rate of Al-Ga-In_3Sn alloy doubled. However, maximum release rate of hydrogen from Zn-containing alloys (1 wt.%, 2 wt.%, 3 wt.%) increased by 2.5, 8 and 6.7 times, respectively. The emergence of Al-Zn micro-galvanic cells in Zn-containing alloys has also contributed to an increase in the hydrogen production rate. Furthermore, at the same temperature, as the amount of Zn doping increases, the maximum hydrogen generation rate is constantly changing and has a maximum at 1 wt.% Zn. If the maximum hydrogen generation rate (almost with high Zn content alloy) is lower than that of the Zn-free alloys, their reaction duration is often extremely long. This is because the eutectic temperature of the GB phase in the high-Zn alloy (more than 1 wt.%) is higher, and more energy is required to promote the hydrolysis reaction.

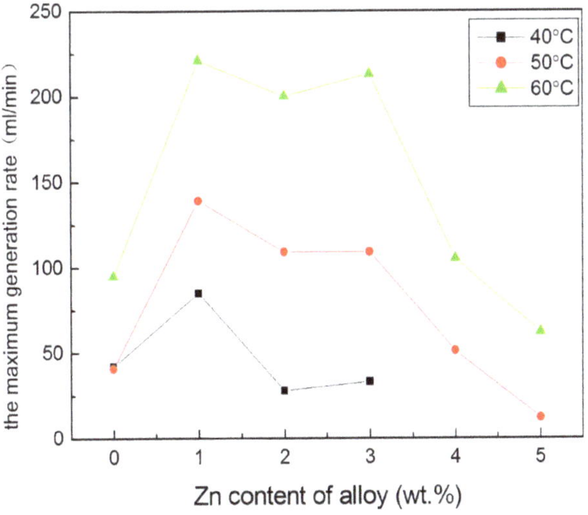

Figure 12. Maximum generation rate versus Zn content.

The factors affecting the hydrogen production performance of different alloys are summarized as follows:

(1) The occurrence of alloy hydrolysis is related to the GB phase and the external surrounding (water bath) temperature.

(2) The GB phase composition of 1 wt.% Zn sample results in its minimum activation energy, and the formation of a small portion of Al-Zn micro-galvanic cells results in the fastest hydrogen production rate.

(3) When the doping amount of Zn reaches 2 wt.%, the area of a single GB phase decreases rapidly. At the same time, when the water temperature is 40 °C, the Al-Zn micro-galvanic cells react slowly, and the local heat generated is small, which delays the arrival of the peak of hydrogen production rate and prolongs the reaction time.

(4) Compared with the 2 wt,% Zn sample, the single GB phase area and the number of GB in 3 wt.% alloy have not changed much, but the appearance of more Al-Zn micro-galvanic cells has increased the local temperature and accelerated the reaction.

(5) When the Zn doping amount reaches 4 wt.% or 5 wt.%, the area and number of GB phases continue to decrease rapidly. The need for higher activation energy means a higher reaction temperature.

In the present work, the mechanisms used to explain the effect of Zn on hydrogen production can be summarized as follows:

(1) The GB phase includes all the eutectic compounds among the grain boundary elements. This is identified to be a key factor that determines whether the Al hydrolysis reaction can proceed. While protecting Al from oxidation, the GB phase can provide a transmission channel to make it easier for Al to enter the reaction site. According to EDX data (Table 2), the GB phase contains a large amount of Al, when the sample is contacted by water, this part of Al first reacts with water. The decrease of the Al concentration will drive other Al molecules from the grains into the GB phase, and make the reaction continue. The addition of Zn will change the composition of the GB phase from Ga-In-Sn to Ga-In-Sn-Zn, which will affect the eutectic temperature and hydrolysis reaction. A small part of In_3Sn was replaced by Zn, which will help increase the relative content of Ga in Ga-In-Sn.and make the mass ratio of the three elements in the raw material close to the eutectic point, but as more low-melting metals (In and Sn) are replaced by Zn (more than 1 wt.%), this will lead to a gradual increase in eutectic temperature.

(2) During the hydrolysis reaction, Al will form micro-galvanic cells with Zn. The release of hydrogen through this electrolytic form is relatively slow, but the heat released helps the hydrolysis reaction. In the alloys, the appearance of the Al-Zn solution and simple substance of Zn will aggravate the formation of micro-galvanic cells. Therefore, with increasing temperature, this variable of hydrogen generation rate is greater for Zn-containing alloys than Zn-free ones.

5. Conclusions

A set of Al-Ga-In_3Sn-Zn alloys was prepared. Through experiments and discussions, the hypothesis that Zn will change the hydrogen production performance of the alloy by entering the GB phase was verified, and it was found that the reaction mechanism cannot only be explained by the eutectic reaction of the GB phase and Al. We can draw the following conclusions:

(1) Zn-doped alloys mainly contain Al (Ga) solid solution, the intermetallic compound In_3Sn, with a high probability of Al (Zn) solid solution. In addition, particles with high Zn content will appear in high-Zn alloys.

(2) In addition to grain size and segregation, the composition of the hydrogen-producing alloy also affects the change in activation energy.

(3) As Zn doping amount increases, the maximum hydrogen production rate always increases first and then decreases. At the same temperature, the reaction time of alloys with high Zn content is longer. At a moderate temperature, Al-Ga-In_3Sn-(3 wt.%)-Zn not only has good hydrogen production

performance, but also has high activation energy, which is conducive to its long-term storage and transportation.

(4) The present work indicates that the alloy hydrolysis reaction is promoted by the GB phase and the Al–Zn micro-galvanic cells.

Author Contributions: Writing—original draft, D.L.; writing—review and editing, Q.G.; formal analysis, D.L.; conceptualization, Q.G.; data curation, Q.A., H.W. and D.L.; investigation, D.L.; resources, J.W.; project administration, C.W. and Q.G. All authors have read and agreed to the published version of the manuscript.

Funding: Financial support for this work was provided by the Industrial Innovation Special Fund Project of Jilin Province (2017C052-1, 2017C045-1).

Acknowledgments: Authors are grateful to editor and referees.

Conflicts of Interest: The authors declare no conflict of interest.

References

1. Ajanovic, A.; Haas, R. Economic prospects and policy framework for hydrogen as fuel in the transport sector. *Energy Policy* **2018**, *123*, 280–288. [CrossRef]
2. Lincoln, S.F. Fossil fuels in the 21st century. *Ambio* **2005**, *34*, 621–627. [CrossRef] [PubMed]
3. Veziroglu, T.N. 21st century's energy: Hydrogen energy system. In *Assessment of Hydrogen Energy for Sustainable Development*; Sheffield, J.W., Sheffield, C., Eds.; Springer: Dordrecht, The Netherlands, 2007; pp. 9–31. [CrossRef]
4. Chi, J.; Yu, H.M. Water electrolysis based on renewable energy for hydrogen production. *Chin. J. Catal.* **2018**, *39*, 390–394. [CrossRef]
5. Secer, A.; Kucet, N.; Faki, E.; Hasanoglu, A. Comparison of co-gasification efficiencies of coal, lignocellulosic biomass and biomass hydrolysate for high yield hydrogen production. *Int. J. Hydrogen Energy* **2018**, *43*, 21269–21278. [CrossRef]
6. Demirbas, A. Comparison of thermochemical conversion processes of biomass to hydrogen-rich gas mixtures. *Energy Sources A Recovery Util. Environ. Eff.* **2016**, *38*, 2971–2976. [CrossRef]
7. Yuan, B.; Tan, S.; Liu, J. Dynamic hydrogen generation phenomenon of aluminum fed liquid phase Ga–In alloy inside NaOH electrolyte. *Int. J. Hydrogen Energy* **2016**, *41*, 1453–1459. [CrossRef]
8. Xu, S.; Liu, J. Metal-based direct hydrogen generation as unconventional high density energy. *Front. Energy* **2018**, *13*, 27–53. [CrossRef]
9. Yang, B.; Zhu, J.; Jiang, T.; Gou, Y.; Hou, X.; Pan, B. Effect of heat treatment on Al Mg Ga In Sn alloy for hydrogen generation through hydrolysis reaction. *Int. J. Hydrogen Energy* **2017**, *42*, 24393–24403. [CrossRef]
10. Wang, Y.-Q.; Gai, W.-Z.; Zhang, X.-Y.; Pan, H.-Y.; Cheng, Z.; Xu, P.; Deng, Z.-Y. Effect of storage environment on hydrogen generation by the reaction of Al with water. *RSC Adv.* **2017**, *7*, 2103–2109. [CrossRef]
11. Tekade, S.P.; Pednekar, A.S.; Jadhav, G.R.; Kalekar, S.E.; Shende, D.Z.; Wasewar, K.L. Hydrogen generation through water splitting reaction using waste aluminum in presence of gallium. *Int. J. Hydrogen Energy* **2019**. [CrossRef]
12. Guan, X.; Zhou, Z.; Luo, P.; Wu, F.; Dong, S. Effects of preparation method on the hydrolytic hydrogen production performance of Al-rich alloys. *J. Alloys Compd.* **2019**, *796*, 210–220. [CrossRef]
13. Guan, X.; Zhou, Z.; Luo, P.; Wu, F.; Dong, S. Hydrogen generation from the reaction of Al-based composites activated by low-melting-point metals/oxides/salts with water. *Energy* **2019**, 188. [CrossRef]
14. Yu, M.; Kim, M.; Yoon, B.; Oh, S.; Nam, D.-H.; Kwon, H. Carbon nanotubes/aluminum composite as a hydrogen source for PEMFC. *Int. J. Hydrogen Energy* **2014**, *39*, 19416–19423. [CrossRef]
15. Liu, Y.; An, J.C.; Jin, W.Z.; Cui, J.; Zhang, W. Insight into the effect of In addition on hydrogen generation behavior and hydrolysis mechanism of Al-based ternary alloys in distilled water—A new active Al-Ga-In hydrolysis hydrogen generation alloy. *Int. J. Energy Res.* **2019**, *43*, 8973–8984. [CrossRef]
16. Godart, P.; Fischman, J.; Seto, K.; Hart, D. Hydrogen production from aluminum-water reactions subject to varied pressures and temperatures. *Int. J. Hydrogen Energy* **2019**, *44*, 11448–11458. [CrossRef]
17. Xu, S.; Zhao, X.; Liu, J. Liquid metal activated aluminum-water reaction for direct hydrogen generation at room temperature. *Renew. Sustain. Energy Rev.* **2018**, *92*, 17–37. [CrossRef]

18. Yang, W.; Zhang, T.; Zhou, J.; Shi, W.; Liu, J.; Cen, K. Experimental study on the effect of low melting point metal additives on hydrogen production in the aluminum–water reaction. *Energy* **2015**, *88*, 537–543. [CrossRef]
19. Wang, W.; Zhao, X.M.; Chen, D.M.; Yang, K. Insight into the reactivity of Al–Ga–In–Sn alloy with water. *Int. J. Hydrogen Energy* **2012**, *37*, 2187–2194. [CrossRef]
20. He, T.; Xiong, Y.; Du, S.; Yuan, Z.; Liang, X.; Huttula, M.; Cao, W. Impact of Li Addition in Al-Rich Alloys on Hydrogen Production in Water. *J. Mater. Eng. Perform.* **2019**, *28*, 2459–2464. [CrossRef]
21. Qiao, D.; Lu, Y.; Tang, Z.; Fan, X.; Wang, T.; Li, T.; Liaw, P.K. The superior hydrogen-generation performance of multi-component Al alloys by the hydrolysis reaction. *Int. J. Hydrogen Energy* **2019**, *44*, 3527–3537. [CrossRef]
22. Wang, F.-Q.; Wang, H.-H.; Wang, J.; Lu, J.; Luo, P.; Chang, Y.; Ma, X.-G.; Dong, S.-J. Effects of low melting point metals (Ga, In, Sn) on hydrolysis properties of aluminum alloys. *Trans. Nonferrous Met. Soc. China* **2016**, *26*, 152–159. [CrossRef]
23. An, Q.; Hu, H.; Li, N.; Liu, D.; Xu, S.; Liu, Z.; Wei, C.; Luo, F.; Xia, M.; Gao, Q. Effects of Bi composition on microstructure and Al-water reactivity of Al-rich alloys with low-In. *Int. J. Hydrogen Energy* **2018**, *43*, 10887–10895. [CrossRef]
24. An, Q.; Hu, H.; Li, N.; Wei, J.; Wei, C.; Wang, H.; Woodall, J.M.; Gao, Q. Strategies for improving flow rate control of hydrogen generated by Al-rich alloys for on-board applications. *Int. J. Hydrogen Energy* **2019**, *44*, 27695–27703. [CrossRef]
25. Wang, H.; Xie, Z.; Chen, C.; Xu, Z.; Dong, S. Ti promotes the reaction of Al alloy with water to rapidly generate hydrogen. *Mater. Res. Express* **2019**, 6. [CrossRef]
26. Wei, C.; Liu, Z.; Wei, J.; Liu, D.; Xu, S.; An, Q.; Xu, S.; Wang, H.; Gao, Q. Effect of alumina on the microstructure and hydrogen production of Al-riched bulk alloys. *Chem. Phys. Lett.* **2020**, 738. [CrossRef]
27. Xie, Z.; Dong, S.; Luo, P.; Wang, H. Enhanced Hydrogen Generation Properties of Al-Ga-In-Sn Alloy in Reaction with Water by Trace Amount of $AlTi_5B$ Additives. *Mater. Trans.* **2017**, *58*, 724–727. [CrossRef]
28. He, T.; Chen, W.; Wang, W.; Ren, F.; Stock, H.-R. Effect of different Cu contents on the microstructure and hydrogen production of Al–Cu-Ga-In-Sn alloys for dissolvable materials. *J. Alloys Compd.* **2020**, 821. [CrossRef]
29. Huang, T.; Gao, Q.; Liu, D.; Xu, S.; Guo, C.; Zou, J.; Wei, C. Preparation of Al-Ga-In-Sn-Bi quinary alloy and its hydrogen production via water splitting. *Int. J. Hydrogen Energy* **2015**, *40*, 2354–2362. [CrossRef]
30. Wang, W.; Chen, D.M.; Yang, K. Investigation on microstructure and hydrogen generation performance of Al-rich alloys. *Int. J. Hydrogen Energy* **2010**, *35*, 12011–12019. [CrossRef]
31. He, T.T.; Wang, W.; Chen, W.; Chen, D.M.; Yang, K. Influence of In and Sn compositions on the reactivity of Al–Ga–In–Sn alloys with water. *Int. J. Hydrogen Energy* **2017**, *42*, 5627–5637. [CrossRef]
32. Du, B.D.; Wang, W.; Chen, W.; Chen, D.M.; Yang, K. Grain refinement and Al-water reactivity of Al Ga In Sn alloys. *Int. J. Hydrogen Energy* **2017**, *42*, 21586–21596. [CrossRef]
33. Kirchheim, R. Grain coarsening inhibited by solute segregation. *Acta Mater.* **2002**, *50*, 413–419. [CrossRef]
34. Dillon, S.J.; Tang, M.; Carter, W.C.; Harmer, M.P. Complexion: A new concept for kinetic engineering in materials science. *Acta Mater.* **2007**, *55*, 6208–6218. [CrossRef]
35. Ahadi, A.; Kalidindi, A.R.; Sakurai, J.; Matsushita, Y.; Tsuchiya, K.; Schuh, C.A. The role of W on the thermal stability of nanocrystalline NiTiWx thin films. *Acta Mater.* **2018**, *142*, 181–192. [CrossRef]
36. Ziebarth, J.T.; Woodall, J.M.; Kramer, R.A.; Choi, G. Liquid phase-enabled reaction of Al–Ga and Al–Ga–In–Sn alloys with water. *Int. J. Hydrogen Energy* **2011**, *36*, 5271–5279. [CrossRef]
37. Wei, C.; Liu, D.; Xu, S.; Cui, T.; An, Q.; Liu, Z.; Gao, Q. Effects of Cu additives on the hydrogen generation performance of Al-rich alloys. *J. Alloys Compd.* **2018**, *738*, 105–110. [CrossRef]
38. Li, L.; Liu, H.; Yan, Y.; Zhu, H.; Fang, H.; Luo, X.; Dai, Y.; Yu, K. Effects of alloying elements on the electrochemical behaviors of Al-Mg-Ga-In based anode alloys. *Int. J. Hydrogen Energy* **2019**, *44*, 12073–12084. [CrossRef]

MDPI
St. Alban-Anlage 66
4052 Basel
Switzerland
Tel. +41 61 683 77 34
Fax +41 61 302 89 18
www.mdpi.com

Crystals Editorial Office
E-mail: crystals@mdpi.com
www.mdpi.com/journal/crystals

www.ingramcontent.com/pod-product-compliance
Lightning Source LLC
LaVergne TN
LVHW070645170726
843515LV00004B/327

* 9 7 8 3 0 3 6 5 0 1 3 8 3 *